MEMOIRS
MUSEUM OF ANTHROPOLOGY, UNIVERSITY OF MICHIGAN
NO. 24

THEDFORD II

A PALEO-INDIAN SITE IN THE AUSABLE RIVER WATERSHED OF SOUTHWESTERN ONTARIO

by
D. Brian Deller
and
Christopher J. Ellis

with a foreword by Henry T. Wright

ANN ARBOR, MICHIGAN
1992

Printed in the United States of America

ISBN 978-0-915703-25-8 (paper)

ISBN 978-1-951538-02-6 (ebook)

The publication of this monograph was supported with a grant from the Ontario Heritage Foundation

Library of Congress Cataloging-in-Publication Data

Deller, D. Brian.

Thedford II : a paleo-Indian site in the Ausable River watershed of Southwestern Ontario / by D. Brian Deller and Chris J. Ellis.

p. cm. — (Memoirs of the Museum of Anthropology, University of Michigan ; no. 24)

Includes bibliographical references.

ISBN 0-915703-25-4

1. Thedford II Site (Ont.) 2. Paleo-Indians—Ontario—Ausable River Watershed—Implements. 3. Stone implements—Ontario—Ausable River Watershed. 4. Ausable River Watershed (Ont.)—Antiquities. I. Ellis, Christopher J. II. Title. III. Title: Thedford two. IV. Title: Thedford 2. V. Series.

GN2.M52 no. 24
[E78.O5]
306 s—dc20
[971.3'2701]

91-25892
CIP

The paper used in this publication meets the requirements of the ANSI Standard Z39.48–1984 (Permanence of Paper)

Table of Contents

Tables

Figures

FOREWORD

Henry T. Wright
University of Michigan

The first late Glacial hunters to move across the St. Clair lowlands and hunt and forage on the east shore of proglacial Lake Algonquin more than 11,000 years ago perceived no international border. The moraines and outwash plains, the spruce parklands grading north into tundra, the harsh rhythm of the seasons at the glacial margin must have presented similar challenges throughout the central Great Lakes. Twentieth-century specialists in the archaeology of these earliest Great Lakes peoples have continued this ancient tradition, no more inhibited by the border than were their earliest predecessors. For two decades or more there has been a free flow of fieldworkers and ideas throughout the region. Researchers from Ontario, Michigan, and Ohio have visited each others' excavations, studied each others' collections, and attended conferences together, to the benefit of all concerned. It is an honor and a pleasure for the University of Michigan Museum of Anthropology to continue this fruitful collaboration between our nations by participating in the publication of this exemplary monograph on an Ontario Paleo-Indian site .

There have been many complementarities in research in the two areas. The sequence of Early Paleo-Indian assemblages first proposed by William Roosa in the 1960s in Michigan (1963, 1965) was greatly amplified by excavations in Ontario in the 1970s. Hypotheses proposed in the 1970s by Brian Deller (1976a) about settlement organization and the 1980s by Chris Ellis about technology (1984) are proving useful research guides for current fieldwork in Michigan. As this monograph demonstrates, our colleagues in Ontario continue to make exciting innovations.

Future progress in understanding the first foragers in the central Great Lakes—and in using the data on their successes and failures in surviving under extreme conditions to evaluate our broader ideas about foraging peoples—depends on the production of monographic site reports with comparable data on artifact assemblages and their excavation contexts. A high standard for such reports was first set by James Fitting's monograph on the Holcombe site in the central Great Lakes (1966) and by George MacDonald's monograph on the Debert site in the Canadian Maritimes (1968). The Vail site monograph by Michael Gramly (1982) and other reports have continued this tradition in New England. However, this volume is the first major site monograph dealing with a Late Glacial site in the central Great Lakes since 1966. It continues the canonical tradition of detailed reporting of variability established in the 1960s, and it has several notable strengths.

—First, thanks to the knowledge of the bedrock sources in the Fossil Hill Formation, gained through the work of Storck and Von Bitter (1989), it is possible to establish the reduction sequence for that major portion of the stone tools fabricated on banded Collingwood Chert. Similar studies on other assemblages will be required to see if the highly patterned core-to-preform-to-tool sequences of Parkhill assemblages typify other Paleo-Indian industries.

—Second, technical and stylistic variability in fluted bifaces has been carefully documented and explicated, with close attention to the limits of the raw materials and to variability introduced by use and resharpening. This builds on approaches set forth in the 1960s by Roosa and continued by all central Great Lakes researchers since that time.

—Third, a new and detailed approach to unifacial tools is proposed, one which combines separate assessments of hafting and edge-use characteristics. This study obviously raises a challenge for future micro-wear studies.

—Fourth, an analysis of the spatial distribution of materials in terms of structured clusters is carefully presented, with the many assumptions inherent in such studies of small samples from plow-disturbed sites clearly explicit.

Future studies must achieve several breakthoughs. One necessity is that of a solidly grounded absolute chronology, without which environmental resource assessments and demographic estimates are shaky at best. The carbon-14 dating of selected botanically identified charcoal fragments , successful at the Whipple site in New England (Haynes et al. 1984), is one approach now being pursued when features are located. The dating of burned chert artifacts with methods such as thermoluminescence is another promising avenue (Simons et al. 1987). Another necessity is the broad application of wear analysis and residue analysis (Hanks 1988) to evaluate the propositions about economic activities and subgroup organization such as those put forth by the authors of this monograph. In addition, the next generations of archaeologists concerned with the earliest Great Lakes peoples will have, without doubt, new questions requiring new research approaches. Fortunately, the carefully excavated Thedford II samples, as well as those from other Paleo-Indian sites in Canada, are well curated and available for future studies.

Acknowledgments

The Thedford II project would not have been possible without the generous assistance of numerous institutions and individuals and we gratefully acknowledge their aid. Responsibility for errors or omissions of content in this report lie solely with the authors. Except for Chapters 2 and 5 which were written mainly by Ellis, this monograph was very much a joint effort and our names appear alphabetically on the title page.

The project consisted of four phases: field survey to locate sites; excavation; analyses; and report production. The field survey was accomplished with financial assistance provided to Deller by the Ontario Heritage Foundation. Field supervision and vehicles were provided by Glenn Stott, Ronald Watts and Reynold Welke. Several individuals provided site information and field assistance. These included: Bill Baxter, Bob Calvert, Ed McLeod, Glen Tedball, Leroy Weed and Frank Wight. Collectively, thanks are offered to the numerous landowners who allowed access to their properties during the survey.

Excavations at Thedford II were financed by grants to Deller from the Ontario Heritage Foundation. Additional funding was provided by Ontario Corporation 418218. Bob Young granted access to his land and allowed us to excavate in his crops. William B. Roosa and Matthew Hill of the Department of Anthropology, University of Waterloo, provided much-needed field equipment. Arthur Roberts carried out aerial photography. During 1981, the full-time field crew consisted of Chris Ellis (field supervisor), Barry Greco (assistant field supervisor), Bryan Cummins, Christine Dodd, Juliet Garfit, Susan Kosowan, Peter MacLean and Andrew Mitchell. Volunteer workers included George Connoy, Bill Fox, Ian Kenyon, Joe Pelly, Darren Webster, Mark Webster, Frank Wight and Stan Wortner. In 1982, the field crew consisted of Juliet Garfit (field supervisor), Pete MacLean, Bud Parker and Melissa White. Volunteers included Mike Austin, George Connoy, Peggy Garfit, Gerry Pelly, Joe Pelly, Darren Webster and Mark Webster.

The analyses were supported by assistance provided to Ellis through teaching assistantships and a Graduate Research Stipend awarded by the Department of Archaeology, Simon Fraser University. As well, the collection of data on the Parkhill site and its incorporation in this report was supported by a Post-Doctoral Fellowship received by Ellis from the Social Sciences and Humanities Research Council of Canada. Deller received assistance from the Department of Anthropology, McGill University, in the form of a McGill University Research Stipend. Christine Dodd labeled and catalogued the artifacts and identified faunal remains recovered from the site. Carl Murphy analyzed the floral materials. At McGill, laboratory assistance was provided to Deller by Sharon Baillie, Katherine Brodeur, Arnie Feast, Maryann Levine and Catherine Timmins.

During the analyses we benefited immensely from the comments and insights provided by Herbert Alexander, Michael Bisson, Robson Bonnichsen, Gordon Dibb, Knut Fladmark, William Fox, Charles Garrad, Richard Michael Gramly, Brian Hayden, Fumiko Ikawa-Smith, Ian Kenyon, James Keron, Jon Lothrop, Moira McCaffrey, Jack Nance, James

Payne, William B. Roosa, Mike Shott, Don Simons, Peter Storck, Peter Timmins, Bruce Trigger and Henry T. Wright III. Bill Fox processed flotation samples at the Ministry of Citizenship and Culture, London (Ontario), and assisted in the C-14 dating. Several individuals allowed us to examine archaeological collections under their care. These included David Keenlyside (Debert site), John Grimes (Bull Brook site), Peter Storck (Fisher and other Ontario sites), Michael Gramly (Nobles Pond and other sites), Don Simons (Gainey site), Mike Shott (Leavitt site), James Payne (Ohio sites), Arthur Spiess and Deborah Brush (Maine sites), William B. Roosa (Parkhill and other Ontario sites), William C. Noble (Ontario finds) and Debra Bodner and William Finlayson of the Museum of Indian Archaeology (Ontario finds).

The preparation of this report was assisted by numerous individuals. The Ministry of Citizenship and Culture, through the auspices of Bill Fox, allowed the use of drafting and darkroom facilities. Ontario Corporation 418218 provided lab space and photocopying assistance. Ellis prepared most of the plates and the figures. Additional figures were drawn by Christine Dodd. Some plates were provided by Peter Storck and William Robertson of the Royal Ontario Museum. Janie Ravenhurst did the artifact drawings. The bibliography was word-processed by Rosemary Ambrose. Ellis would like to express special thanks to Morley and Queenie Deller for all of their hospitality during extended sojurns in Mt. Brydges. We would like to especially thank James V. Wright, L.J. Jackson and Michael Gramly for their detailed comments on earlier drafts.

Finally, we would like to acknowledge several individuals who provided consistent support and encouragement. Deserving of special mention are Herbert Alexander, William Fox, Brian Hayden, W.V.V. Pardy, Don Simons, Bruce Trigger, Henry Wright and William B. Roosa.

CHAPTER 1

Introduction

This monograph describes and analyzes data recovered from Thedford II, a Paleo-Indian, fluted point associated site in southwestern Ontario. The site is only one of a series of Early Paleo-Indian sites located by Deller (1976a, 1979, 1980a, 1980b; Deller et al. 1986) near the Thedford embayment occupied by pro-glacial Lake Algonquin. Most of these sites, including Thedford II, can be assigned to the Parkhill complex or industry, first defined by Roosa (1977a, 1977b; Wright and Roosa 1966) and subsequently elaborated upon by several investigators (e.g. Deller 1980a, 1988; Ellis 1984; Payne 1982; Roosa and Deller 1982; Storck 1983, 1984a; Wright 1981a). Diagnostic of this industry are Barnes points (Roosa 1965). Other Parkhill industry sites in the Thedford area include McLeod (Ellis 1979, 1984), Dixon (Deller 1980a; Ellis 1984) and Parkhill (Roosa 1977a, 1977b). Research at Thedford II, in combination with data from these other sites, provides opportunities that currently are rare in the Northeast to document and explain variation in a series of Paleo-Indian sites that are closely related in time and space.

To provide background data for the descriptions and analyses that follow, this introductory chapter will discuss the site's location, layout and setting, a brief history of site investigations, the excavation techniques employed, and a brief discussion of the probable age of the site and its environmental context at the time of the Paleo-Indian occupation. In subsequent chapters, we will discuss: lithic raw materials and primary manufacturing procedures; bifacial tools; unifacial tools; flaking debris; features; spatial distributions; and the place of Thedford II within the broader context of early North American occupations.

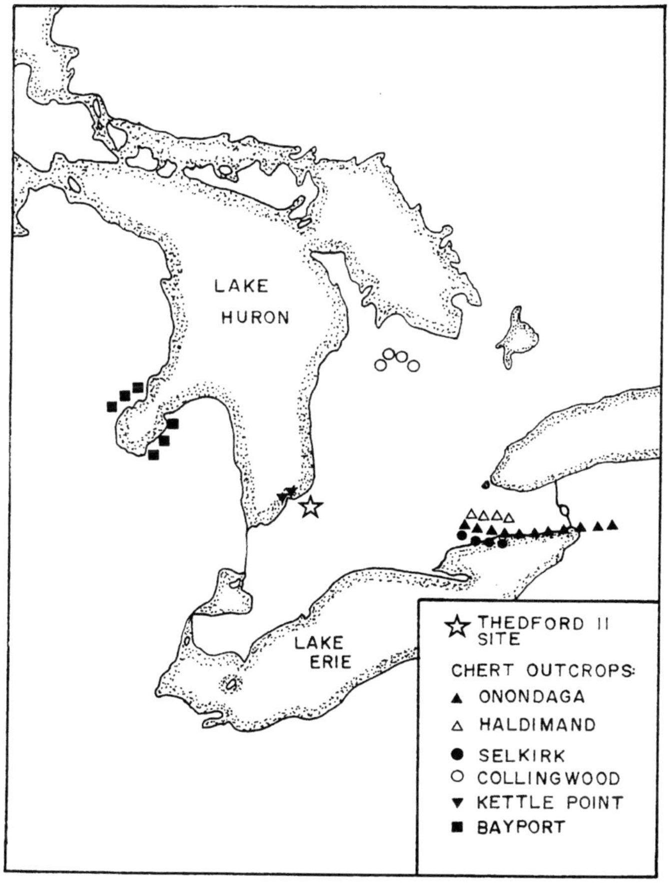

Figure 1. Thedford II site and chert outcrop locations.

Site Location and Setting

The Thedford II site is located in Bosanquet Township, Lambton County, Ontario (Figs. 1 and 2). Although the surrounding terrain exhibits a somewhat rolling topography in contrast to other Paleo-Indian sites in the area such as Parkhill, the site is on a relatively flat terrace at an elevation of about 194 m a.s.l. Surface deposits on the terrace are composed of glacio-

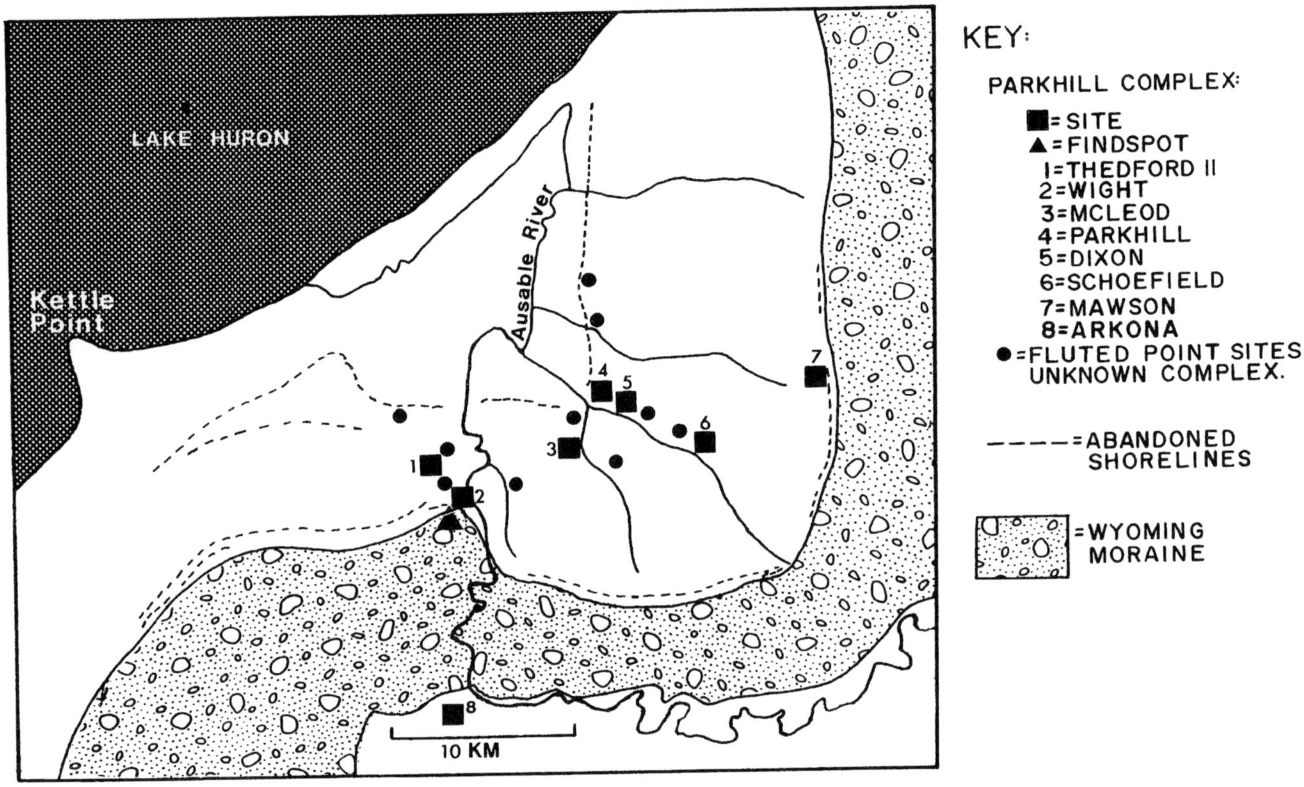

Figure 2. Paleo-Indian sites and finds in the Thedford/Parkhill area.

lacustrine deep water deposits probably laid down by Lake(s) Warren (see below and Cooper 1979). About 750 m to the west of the site the land begins to rise rather abruptly, eventually forming a large, long, northwest to southeast trending ridge (elevation between 213 and 221 m). The top of this ridge represents the only exposed surface bedrock in the immediate area. The lower flanks of this feature adjacent to the site are covered with till deposits laid down during the Port Huron stadial ice advance around 13,000 B.P. (Cooper 1979:23).

The site overlooks a ravine on the south which presently contains a low marshy to wooded area fed by local springs (Fig. 3). This ravine cuts north 80 m east of the site and effectively separates it from the Ausable River which flows in a northerly direction about 1100 m to the east. The site is bordered on the north by a shallow, wooded, depression which deepens to the east and joins the ravine noted above. This ravine then continues to the northeast and joins several others before opening up on the main Ausable River plain (the Thedford embayment) about 1 km northeast of the site.

The Thedford II site is not located on any traceable pro- or post-glacial lake strandline but at least three of these features are located in the area. The nearest is the pro-glacial Lake Warren strandline(s) which straddles the north edge of the Wyoming moraine about 1250 m south of the site (Fig. 2). As with surficial till deposits in the area, this moraine is a product of the Port Huron ice advance. Lake Warren drained, exposing the site area, around 12,500 B.P. (Cooper 1979:39; Fullerton 1980).

The second strandline in the region is traceable in an east-west line to a point about 6.5 km northwest of Thedford II at an elevation of around 195 meters. This strandline has been variously attributed to either Lakes Lundy or Grassmere which briefly formed and drained between 12,500 and 12,400 B.P. (Cooper 1979; Eschman and Karrow 1985; Fullerton 1980). The waters associated with this strandline may have approached the immediate vicinity of the site which is just at or below its elevation. However, definite evidence of this has not been found and certainly, this strandline cannot be traced through the site area today.

Finally, the site is landward of the abandoned shoreline of post-glacial Lake Nipissing Phase I (ca. 5500–4500 B.P.; Lewis 1969; Larsen 1984). This strandline is trace-

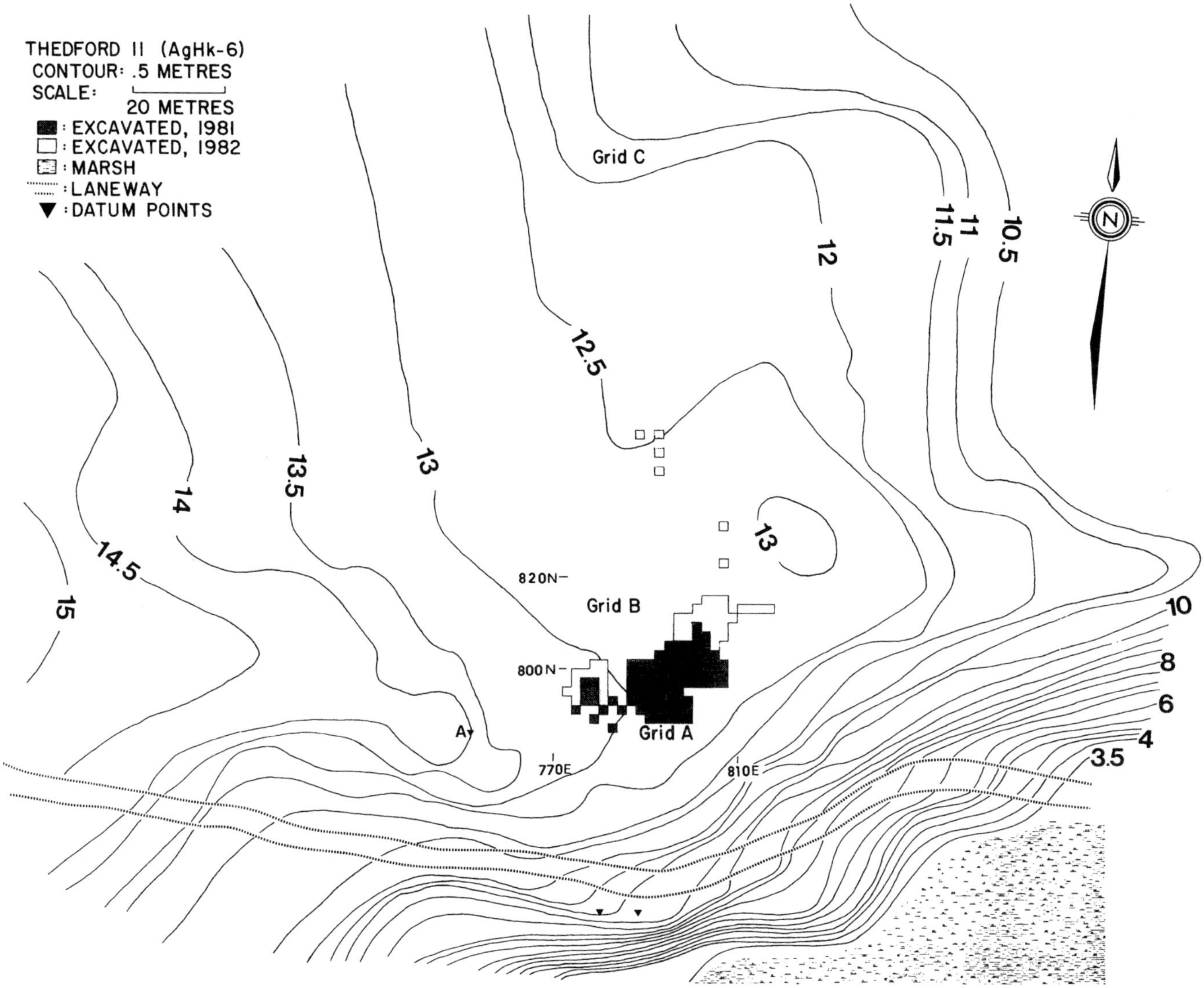

Figure 3. Topographic map of the Thedford II site. Datum A assigned arbitrary elevation of 14 meters.

able in an east-west line about 3 km north of Thedford II but disappears in the Ausable River mouth only to reappear east of the Ausable about 3.7 km northeast of the Thedford II site. This strandline is often referred to as Nipissing-Algonquin (e.g. Cooper 1979), denoting the fact that pro-glacial Lake Algonquin (ca. 12,000 to 10,400 B.P.) is suspected to have reached a similar elevation (184 m) in the study area (Hough 1963; Karrow et al. 1975). However, as Karrow (1980) has recently noted, the Lake Algonquin beach is only inferred to have been present at this location as the later "Nipissing transgression extensively removed and destroyed the older Algonquin features" (Karrow 1980:1271). Geological investigations at the Parkhill site have suggested that the Algonquin strandline was present in the area but its exact elevation is difficult to resolve with present evidence.

The site's location and relationship to Lake Algonquin is particularly important since the time of this lake's existence corresponds to the probable age of the site (see below). However, as a result of the Nipissing transgression, this relationship is difficult to determine. It is probable that an inlet of Algonquin (and certainly, one of Nipissing; Kenyon 1980) existed at the Ausable

River mouth. Even assuming the Algonquin strandline was in the same position as that of the Nipissing phase, the fact the latter strandline cannot be traced in the river mouth area obviates inferences as to the exact site location relative to Lake Algonquin. On present evidence, it is possible that: (1) the Algonquin shoreline was a kilometer or more northeast of the site; (2) the ravines surrounding the site were small, shallow, flooded "fingers" or inlets in an Algonquin bay at the river mouth; or (3) these small ravines were marshy areas bordering an Algonquin bay. In short, Lake Algonquin could have been anywhere from near the site to a distance of a kilometer or more away. Resolution of this problem demands extensive geological assessment of the deposition and sedimentation history in the Ausable River area.

Site Layout and Investigations

The Thedford II site was first discovered and reported by Deller (1980a:17–25) during a reconnaissance for Paleo-Indian sites in June, 1978. The site was discovered following assumptions that: (1) the Thedford embayment, when occupied by Lake Algonquin, provided a natural obstacle that had to be circumvented by caribou migrating northward out of southwestern Ontario; (2) this feature served to concentrate caribou migration routes at its southeastern extremity; and (3) this concentration of game would have attracted early Paleo-Indian hunters to the area. Based largely on these conjectures, a decision was made to survey for sites at strategic locations in the corridor where caribou could be most logically intercepted. Canvassing of local residents revealed Paleo-Indian artifacts from the same field in two local collections. Subsequently, two point fragments from another collection were found to fit together with point fragments recovered in the excavations indicating they were also from the site. Originally, these two fragments (see Deller 1979: Fig. 4:8a, 8b) were believed to be from a different locality.

Initial surface collection delimited two areas of Paleo-Indian lithic debris. In addition, artifacts representing later prehistoric occupations were found across the site area. Within the main area excavated, diagnostics are primarily small, well-made points (Fig. 4) which are attributable to occupation(s) during Terminal Archaic times by groups representing the "small point" tradition of 3500 to 3000 B.P. (see Deller et al. 1986; Spence and Fox 1986). These items are largely on lithic materials not employed by Paleo-Indian knappers at the site (see next chapter) and thus, can be easily distinguished from the Paleo-Indian debris.

The northernmost concentration of Paleo-Indian material (Grid C) is located near the shallow depression (Fig. 3). Apparently, it is quite small, having yielded only a few small retouch flakes, a fragment of a channel flake and an end scraper bit. The second area is much more extensive. It is located between and north of two low knolls at the south end of the site. It was clear that this area consisted of at least five concentrations in relatively close juxtaposition. One of these appeared somewhat discrete and was referred to as Grid B while the remainder were grouped together as Grid A. However, subsequent research indicates Grid B is not as discrete as originally thought (see Chapter 7) and it can be interpreted as simply an extension of Grid A. Within Grid A, surface collections suggested four concentrations referred to as *A-west*, *A-centre*, *A-east* and *A-southeast*. Subsequent excavations located another unsuspected concentration named *A-northeast* (Fig. 5). The earliest surface finds were recorded on sketch maps in relation to bench marks and topographic features. Nonetheless, many of these finds can be assigned to particular concentrations within the area. The location of subsequent finds was precisely recorded in relation to permanent data points.

Three aspects of recent site history may affect its interpretation. First, there has been a small amount of bulldozing to construct the laneway between Grid A and the ravine (Fig. 3). This probably had no effect on the Paleo-Indian areas of the site as only more recent artifacts, such as Late Woodland points (Fig. 4L), have been found in the lane area. Second, the site has been plowed, which certainly has an effect on interpretations of intra-area artifact distributions. The west half of the site has been cultivated for a century or more. The eastern half was not cleared and cultivated until 20–25 years ago. During excavations, the old fence line which separated the two halves was located (Fig. 5). Finally, a buried tile drain constructed by the landowner was encountered in A-northeast. This alteration appears to have had some effect on artifact distributions in the area as will be discussed in later sections.

Excavations

Excavations were carried out in 1981 and 1982. Excluding six test squares (24 square meters) excavated just to the north of Grids A and B, which were sterile in terms of Paleo-Indian material, excavations were confined to Grid A proper. In all, 500 square meters were excavated using a grid of two-meter squares. The north-south axis of the grid was oriented to magnetic north

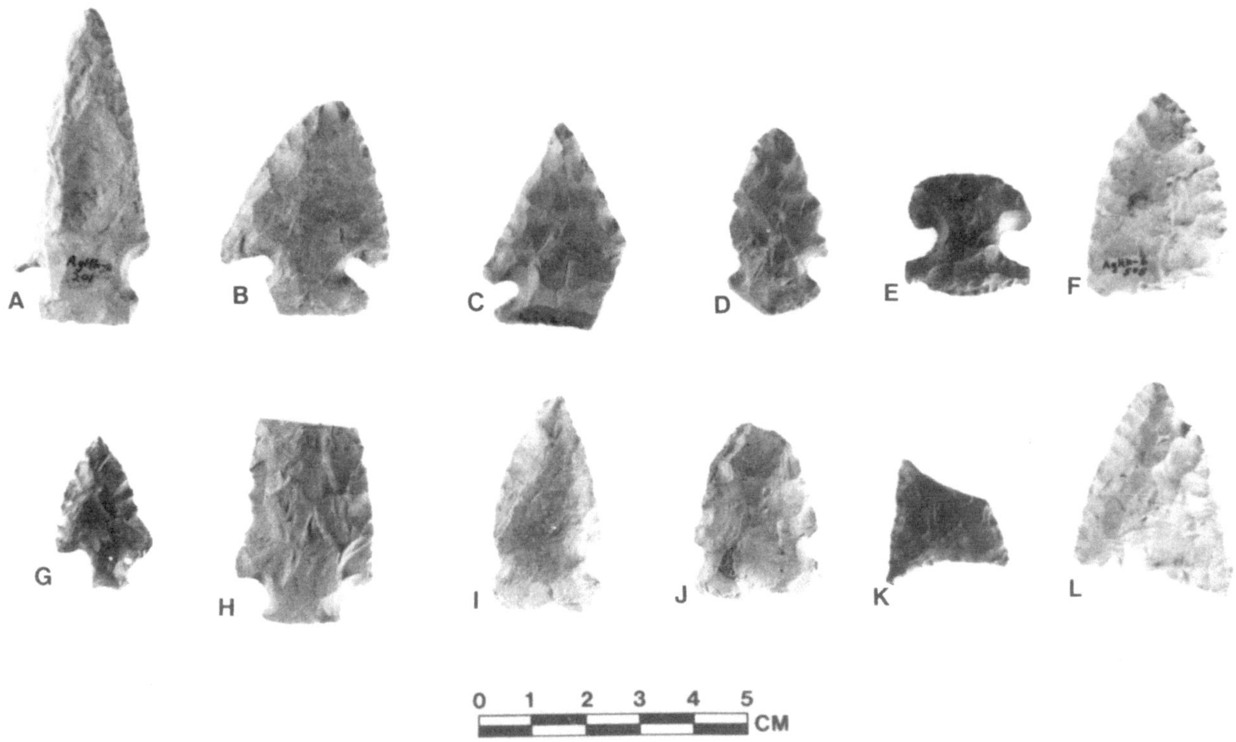

Figure 4. Notched and triangular Points, Thedford II site.

as of 1981. This also approximated the direction of site plowing. The two-meter grid squares were referred to by the intersection coordinates of the north-south and east-west baselines at their southwest corners. The four one-meter quadrants or subsquares within each main unit were numbered counterclockwise from 1 to 4 with the southwest quadrant being number 1.

The initial baselines were established with a transit. Their intersection point was arbitrarily designated point 800 North/ 800 East and was located between two of the Grid A surface concentrations (A-centre and A-east). A series of two-meter squares was then triangulated within each of these two concentrations and excavations commenced (Fig. 6).

In 1981, the initial squares were opened up in a checkerboard pattern. It was decided to continue this pattern until either: (1) the limits of the concentrations were reached or (2), there was suitable time left in the field season only to excavate the intervening squares in the existing checkerboard pattern. This latter alternative was necessitated so that continuously excavated areas suitable for intra-site spatial analyses would be available. Although the limits of some concentrations were fairly effectively delineated in certain directions, it was clear that other areas were not fully exposed. In order to partially overcome this problem, more limited excavations were carried out in 1982. While this served to define more accurately concentration limits, some undoubtedly extend beyond the excavated areas. In addition, two of the concentrations (A-southeast and B) are known only through surface collection.

Squares were excavated in the following manner. First, the plow zone from each one meter subsquare was removed and screened. Only one shovelful of dirt was screened at a time so that tools found in that shovel-load could be located precisely. Coordinates were taken to the center of the shovel-load location for all tools, channel flakes and uniface retouch flakes (see Chapter 5) noted in the field. Thus, plow zone finds are noted to an area of around 25 by 25 cm (i.e., the area of

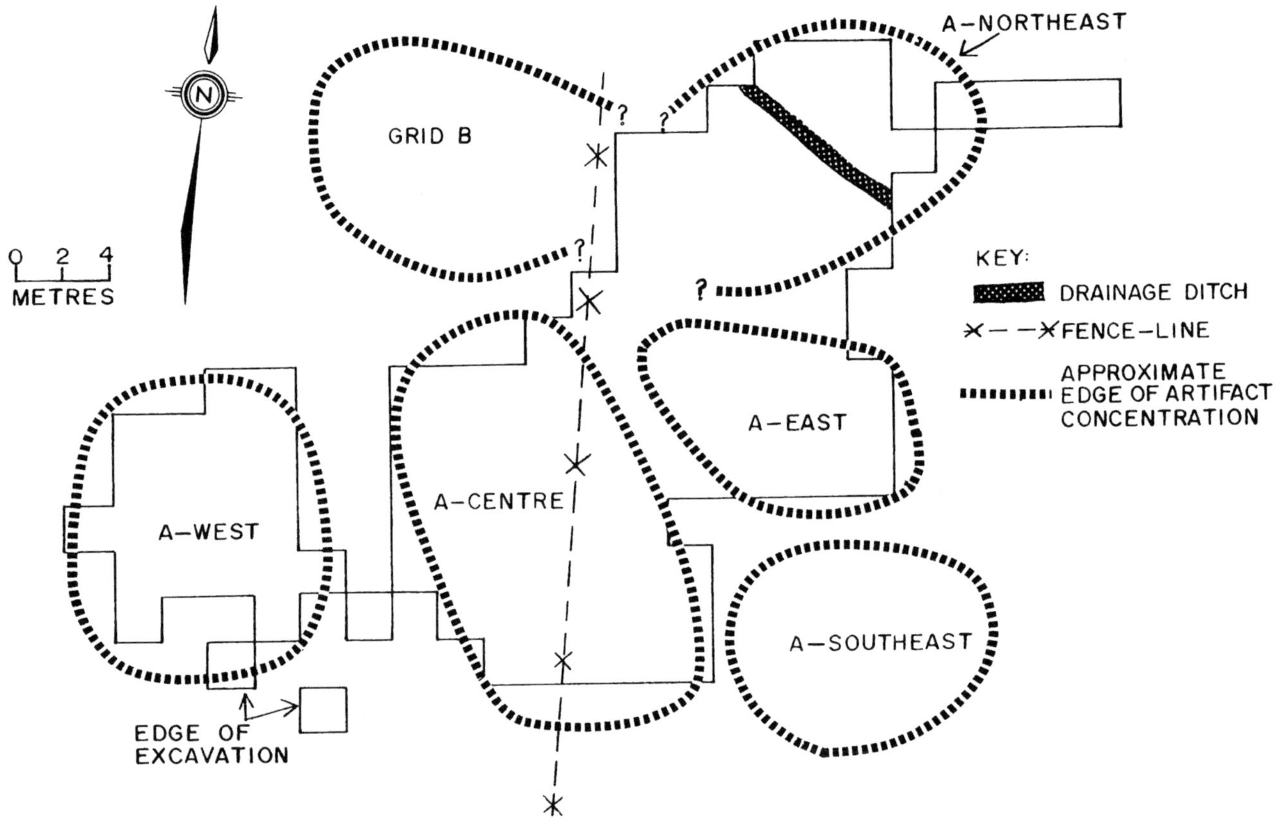

Figure 5. Location of site areas, Grids A and B.

the shovel) except for most flaking debris which was bagged by subsquare. Although time consuming, piece-plotting the data was advantageous in several respects: (1) we have exact data on the separation of refitted fragments which is useful in estimating the effects of plowing on spatial distributions; (2) the edges of the concentrations are more clearly and unambiguously delineated; and (3) statistical analyses of the spatial distributions carried out by Greco (1985:152) indicate that some patterning could be detected in the data using piece-plotted data whereas such patterning was not evident where other techniques using simple presence-absence by excavated unit were employed.

In each square, the northeast quadrant (3) was passed through 1/8" mesh whereas 1/4" was used in all other subsquares. This technique allowed the collection of a sample of the much smaller materials from the site and at the same time avoided the time-consuming work needed to screen all soil through the smaller mesh size. Moreover, it ensured that all 2 by 2 meter units were excavated in the same manner and could be used as comparable units across the site for plotting density distributions of items such as waste flakes. Baulks ten centimeters wide were left between some units to retain back dirt, prevent wall collapse and to serve as guides for subsoil arbitrary level excavation. These were removed and sieved if there was evidence of dense occupation in an area.

Once the plow zone was removed, the plow zone-subsoil interface was cleaned off by troweling (and then screened) in order to expose possible subsoil cultural features. In the subsoil, the surface was continually examined and re-examined for features or cultural material. Plan scale drawings and photographs were made at interface of all cultural and non-cultural features. All of the latter (if they could be recognized as such in plan view) were removed and screened separately prior to taking down the subsoil any further. Cultural pit fea-

Figure 6. Excavations in A-centre (foreground) and A-east (background) looking southeast, July 1981.

tures (e.g., those with a visible outline) were carefully excavated with fine tools.

Excluding some artifact features (e.g., subsoil concentrations of artifacts which were not in pits visible to the naked eye), all cultural features were sampled for flotation. Generally, just under 50 percent of the feature volume was taken for flotation. These samples were processed using a S.M.A.P. flotation device. The light fraction was recovered in 6.3 and .59 mm sieves while the heavy fraction was collected in 2 mm mesh.

Subsoil excavations proceeded by 5 cm arbitrary levels contoured to fit the profile of the plow zone-subsoil interface in each square. Unless the cleaning of the interface or the excavation of previous subsoil levels suggested *in situ* subsoil cultural material, these levels were carefully removed by shovel-shining and screening. Flaking debris was bagged by arbitrary level and subsquare in these cases. If subsoil cultural material was encountered, fine tools and screening were used. Subsoil was screened through the mesh size appropriate to the subsquare, as in the plow zone. Subsoil materials found in place, including flaking debris, were plotted on maps. Three dimensional coordinates were taken on *in situ* tools and much of the flaking debris, horizontally with reference to grid pins and vertically with reference to a central vertical datum point (the ground surface at Datum A; Fig. 3). All squares were excavated at least 15 cm into the subsoil. If subsoil materials were found, the square was taken down until at least one sterile 5 cm level was excavated.

Age and Environment

Materials suitable for radiometric dating of the Thedford II Paleo-Indian occupation were not recovered. However, C-14 dates from other sites in the east in good association with artifactual materials suggest an occupation between roughly 11,000 and 10,000 B.P. (see Haynes et al. 1984 for a recent summary). In fact, these dates suggest the Early Paleo-Indian occupation had ended sometime prior to 10,000 B.P. and perhaps as early as 10,400 B.P. (Ellis 1984; Goodyear 1982a: 329).

In the Great Lakes region, geochronological estimates (see Jackson 1983; A. Roberts 1984) also suggest a termination age prior to 10,000 B.P. In particular, fluted point materials in reliable contexts have not been found on the bed of pro-glacial Lake Algonquin which drained around 10,400 B.P. (Karrow et al. 1975). There are, however, two major reservations which must be expressed with regard to these geochronological estimates.

First, less attention has been directed towards surveying areas below this strandline than to landward areas. Second, it is possible that the Nipissing transgression in the Lake Huron basin obliterated or obscured pre-Nipissing sites in this particular area. On the other hand, it must be noted that some early sites such as Tedball and one area on the Heaman site (Deller et al. 1986; Deller 1976b; Ellis and Deller 1986) are located below the Nipissing-inferred Algonquin strandline. Both sites have yielded concentrations of tools and flaking debris tumbled and patinated by Nipissing Phase I action. Both have yielded Late Paleo-Indian materials (either Holcombe or "plano" forms; see Fig. 31L). Thus, if fluted point sites exist below the strandline of Nipissing-inferred Algonquin, they should be just as easy to locate as these slightly later manifestations. However, to date, no fluted point sites have been reported in these areas. Moreover, Kettle Point chert from just to the northwest of Thedford II (see Fig. 3) was rarely used by Early Paleo-Indian knappers. Since this source was probably within the area inundated by Algonquin, its rarity also suggests the Early Paleo-Indian occupation was contemporary with that lake (see Deller 1989).

If we accept that most fluted point sites date between 11,000 and 10,400 B.P., it is possible to infer environmental associations. The Thedford II project did not recover any substantial data in this regard. However, considerable information was recovered nearby during the investigations at the Parkhill site. The report on that site is currently in preparation and we will reserve detailed discussion of the environment for that report. Here we will confine ourselves to a brief overview of possible environmental associations.

The suggested time range indicates contemporaneity with two widely recognized pollen zones: one dominated by spruce prior to 10,600 B.P. and one dominated by pine pollen after that date (Karrow et al. 1975:53; McAndrews 1981). The earlier zone is characterized as an open tundra parkland or woodland but it is clear that it varied from modern high latitude environments of this nature. Deciduous trees such as oak were present in small numbers in well-drained, upland localities (Karrow et al. 1975:57). As well, there are some isolated localities along the glacially fed lakes which provide evidence of more arctic-alpine climates (Ashworth 1977:1633). The later pine zone suggests the appearance of warmer and drier climates and a more closed "boreal forest" although, in contrast to modern boreal forests, there is a relatively high frequency of hardwoods (Mott and Farley-Gill 1978:1109). Given that Parkhill industry sites such as Thedford II represent the second of three time-sequential Early Paleo-Indian industries in the Great Lakes area (see below), it has been suggested that this complex coincides largely with the transition between the two major zones about 10,600 B.P. and that more northern sites such as Thedford II and Parkhill were located in a spruce-dominated vegetation cover (Deller and Ellis 1988).

These vegetation covers probably supported a variety of faunal species suitable for exploitation by Paleo-Indians. However, exact associations with the pollen zones and the contemporaneity of all species cannot be assumed other than perhaps, an association of mastodons with the earlier spruce parkland (Dreimanis 1968; McAndrews and Jackson 1988; Winn 1977). Caribou, elk, musk-ox, mammoth, deer, moose, giant beaver, ground sloth, snowshoe hare, ground squirrel, fox and beaver (*Castor canadensis*) are just some of the species which might have been present either throughout or for limited times during the fluted point associated occupations of the area (Jackson 1978; Brown and Cleland 1968; Cleland 1966:19–20). Of these, there is definite evidence that elk (Ogden 1977:10), beaver (Spiess et al. 1985) and caribou (Cleland 1965; Spiess et al. 1985) were exploited in other areas of the Northeast. Recently, *Leporidae* (hare/rabbit), ?*Canidae* (fox), and *Cervidae* (caribou/deer), have been identified from the Udora site in south-central Ontario (Storck 1988).

At present, it is debatable whether the caribou were barren-ground or some other subspecies, or even if modern subspecies had evolved or exhibited comparable behaviours to modern forms (Simons et al. 1984:267). Certainly, the environments were not directly comparable to those inhabited by modern forms such as barren-ground caribou. It is our opinion, however, that the nature of the Paleo-Indian site record throughout the Northeast is indicative of at least some reliance on communal hunting (see arguments in Deller and Ellis 1988) as one aspect of subsistence, and caribou seems a logical choice as an object of this pursuit.

Fishing is a possible subsistence practice because fish species are thought to have been present in Lake Algonquin (Karrow et al. 1975:65). It is the opinion of one of us (Ellis) that fishing was probably not as important a subsistence pursuit as it became in later times (see also Cleland 1982). In particular, the vast and rapid changes in water levels, flow, turbidity, and lack of nutrients associated with the formation of the Great Lakes may have inhibited the ability of these species to become well-established or to provide a consistent, reliable, food source. However, this reliability may have increased through time. This is largely speculation.

Finally, although these early environments are not totally similar to modern spruce parklands or boreal forests, it seems safe to assume that plant foods did not play a major role in the diet of Paleo-Indians in the area.

CHAPTER 2

Lithic Raw Materials and Primary Reduction

Lithic Raw Materials

Chemical, structural, and detailed visual analyses demonstrate that the vast majority of the tools on Parkhill industry sites in the Thedford/Parkhill area northeast into south-central Ontario are on Collingwood chert (Deller 1979; Ellis 1984:42–47; Laye 1977; Arthur Roberts, pers. comm.; Peter Sheppard: pers. comm.; William Fox: pers. comm.; Storck and von Bitter 1981, 1989). At Thedford II and other sites in the area such as Parkhill (Ellis 1979; Deller 1980a; Roosa 1977a, 1977b), McLeod (Ellis 1979, 1984) and Dixon (Fig. 2), this material typically accounts for 75% or more of the assemblages whereas it is totally lacking on sites of more recent ages. As a result, it is diagnostic of Paleo-Indian sites in southwestern Ontario and, given its visual distinctiveness, it has been of immense value in recognizing early sites from surface collections that otherwise lack diagnostic implements (see, for example, Deller and Ellis 1984).

The focus on a single variety of chert often quarried at a distant source, has been noted at a significant number of Great Lakes Paleo-Indian sites such as Barnes (Voss 1977:225), Gainey (Simons et al. 1984:267), Crowfield (Deller and Ellis 1984:43), Potts (Gramly and Lothrop 1984:129) and Holcombe (Fitting et al. 1966:126). At these sites, the material can be from sources up to 350 km distant. This pattern probably occurs in other areas of the East (see Meltzer 1984, 1985) such as at the Shoop site (Witthoft 1952:470–71) and the Debert site (MacDonald 1968).

The remainder of the Thedford II Paleo-Indian assemblage is primarily on Bayport chert although one tool on Onondaga and two tools on unknown materials were also recovered. In addition, large numbers of Kettle Point chert artifacts were recovered from the site. However, these are not Paleo-Indian associated as the only diagnostics on this material are of an Archaic/Woodland affiliation and this material concentrates spatially in the area of pit features associated with later site occupations as well as their diagnostics. It should be noted that some of the tools associated with the later occupations are also on Bayport chert. This creates some problems when sorting out waste and tool fragments associated with the Paleo-Indian occupation as will be discussed in more detail later.

In the following, Collingwood and Bayport chert are described. These descriptions will serve to indicate some of the detailed visual similarities between the artifact and outcrop samples. More importantly, in the case of Collingwood chert, the description will serve to emphasize several characteristics of the chert useful in constructing models of lithic manufacturing procedures.

Collingwood Chert

This chert originates in the Fossil Hill formation of the Collingwood, Ontario, area about 175–180 km northeast of the site. In the late 1970s, research by Peter Sheppard (1977) and William Fox (pers. comm.) succeeded in locating till sources in that area. More recently, a concerted research program by P. von Bitter of the Royal Ontario Museum, in cooperation with Peter Storck (Storck 1984a, 1984b; Storck and von Bitter 1981, 1989; Eley and Von Bitter 1985), has succeeded in locating bedrock outcrops of the material. These studies have consistently indicated that the material in outcrop and concentrated till deposits is restricted to the northern area of the Niagara escarpment in the Collingwood-Owen Sound area.

Collingwood chert occurs only in a bedded as opposed to nodular form. At surface outcrops, the chert occurs in up to three overlying beds or bands (Storck and von Bitter 1989). The blocks of the chert from till and outcrop sources often exhibit a dolomite cortex at the juncture of the old chert bed with the surrounding matrix. This juncture is well defined and forms relatively straight, regular margins on both archaeological and lithic source samples. It should be emphasized that when cortex and other original surfaces occur on the archaeological materials from Thedford II, they do not exhibit the tumbled or rounded surfaces characteristic of secondary deposits such as those in stream beds or glacial till. In other words, the material undoubtedly was obtained from actual outcrops.

The chert is opaque to slightly translucent. Generally, it is fine grained and often exhibits small speckles or pits stained black by iron oxides. It has few macrofossils, and is of a medium to high lustre. Often, the high lustre on archaeological materials appears to be a product of heating as it co-occurs with features such as potlid scars. The artifact and source samples range in color from a very pale brown (10YR8/3) to beige (10YR8/2) to a white/light grey (N8; N7) in the Munsell system.

An important aspect of the chert which will figure in the artifact analyses is that it is usually banded. The *surfaces* of the banding planes parallel the cortical surface while the *edges* of these planes form straight lines parallel to one another when the bed or band of the chert is viewed in a vertical section encased in its bedrock matrix (Fig. 7). Not all of the chert is banded. One particular variety which is commonly not banded is a "soft" chert with a mottled surface appearance. This variant may be a peculiarity of a particular outcrop or alternatively, may be due to variation between separate beds of the chert at a single outcrop. So far, this variety is known only from till sources in the bedrock areas.

The artifacts in the assemblage exhibit two other attributes in common with source samples. These attributes reflect weathering at the source since they have been partially removed by retouch on the artifact samples. First, there are *natural fracture planes,* which are randomly oriented with respect to the original chert bed and cut through pieces of the material rather than being exposed on artifact or source samples. When exposed by flaking, these fracture planes are stained red due to iron content. Second, there are *weathered surfaces.* These are glossy to dull and "varnish-like" in appearance. Unlike the natural fracture planes, these surfaces are never stained red and never cut through artifact or source samples. Rather, they always appear on unflaked dorsal surfaces of artifacts and exposed surfaces of bedrock samples.

Finally, it should be noted that the chert artifacts have been slightly weathered or "yellowed" such that flake scar surfaces appear dull when compared to fresh breaks and recently flaked surfaces from outcrops. This facilitates the recognition of recent breaks or modifications in the artifact sample.

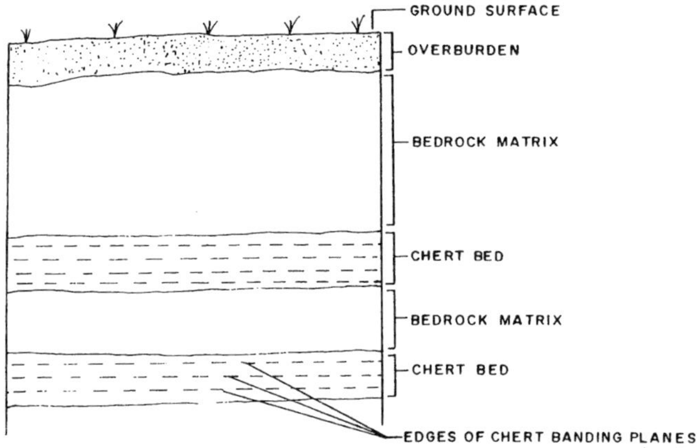

Figure 7. Schematic vertical section of Collingwood chert in its bedrock matrix.

Bayport Chert

This chert originates in the Upper Mississippian Bayport formation in a dolomitic to sandy limestone. Several studies dealing with its identification and distribution have been published (Dustin 1935; Luedtke 1976; Fitting et al. 1966; Ozker 1976). Outcrops of Bayport chert are restricted almost entirely to the Saginaw Bay area of Michigan, about 120–150 km northwest of Thedford II. Sources are known both above and below the inferred Algonquin strandline (Luedtke 1976: Fig. 10; D. Simons: pers. comm.; Voss 1977:255).

Although Bayport chert can occur in a bedded form, it usually occurs as nodules. The most distinctive characteristic of this chert is a concentric banding which begins at the center of nodules. Towards the exterior of the nodule, the chert tends to become "browner" indicating a higher limestone content. Otherwise, the chert ranges in colour from a light grey (10YR7/2) to a light brownish grey (10YR6/2) to a dark grey (N6 to N4) in colour. The chert is usually highly fossiliferous with both artifact and source samples exhibiting the occasional large shell fragment. It is opaque, can be somewhat coarse to fine grained in texture, has a dull to medium lustre and exhibits small quartz crystal grains which sparkle when rotated under a light source. One Bayport chert tool in the Thedford II assemblage exhibits a cortical surface which suggests the use of an outcrop rather than a secondary source.

As with Collingwood chert, it is relatively easy to recognize recent breaks on Bayport chert items, as these appear fresh when compared to older breaks and purposefully flaked surfaces on artifacts.

Primary Reduction

Primary stages of lithic manufacture (core preparation and tool blank production) were not carried out at Thedford II nor on other Parkhill industry sites in the area (see Ellis 1979 and Chapter 5). These activities apparently were restricted to locations near the chert outcrops. Nonetheless, significant information on the early stages of manufacture can be derived from an examination of blanks that have been retouched into tools. Since this information has been detailed elsewhere (Ellis 1984:61–130), only a brief overview is presented here. An understanding of the early stages of tool production is important because it often is of diagnostic value and also, as will become clear below, it is basic to understanding the variation within the assemblage as a whole and even within particular tool types. Appendix A explains most of the terminology employed and the methodology used to measure blank attributes.

"Tops and Sides"

The available data indicate the reduction sequence began with a large "quarry block" which approximated a section of the chert bed. These blocks exhibit two kinds of surfaces. First, there are "top" surfaces (Fig. 8a) which approximate the juncture of the chert bed with its surrounding limestone matrix. Therefore, these surfaces can be cortical. These surfaces also parallel the surfaces of the banding planes in the chert. Those surfaces paralleling the banding plane surfaces are referred to as "tops" whether they are cortical or not.

The second category of surfaces occur at right angles to "top" surfaces. They approximate *in orientation* those surfaces of the chert bed which would be exposed if the bed was viewed in vertical section still encased in its bedrock matrix (as on Fig. 7). These surfaces are also at right angles to the banding plane surfaces in the chert so that the edges of the bands appear as straight lines paralleling one another (Figs. 7, 8a). All surfaces at right angles to the top surfaces exhibiting such straight, parallel bands are referred to as "sides" (Fig. 8a), *whether or not* they are original surfaces of the quarry block or subsequently flaked surfaces.

For simplification purposes, the quarry blocks are illustrated as squarish on Figure 8; that is, all the top and side surfaces meet each other at approximately 90°. While examination of such blocks from outcrop samples provided by Charles Garrad indicates the blocks consistently exhibit about 90° junctures between top and side surfaces, this is not the case with the junctures of adjacent side surfaces, the angle of which can vary considerably (Fig. 9). As a result, the initial blocks can exhibit more or less than the four "sides" implied on the simplified diagrams.

Systematic Reduction

After obtaining a quarry block, the evidence indicates Paleo-Indian knappers reduced it systematically rather than in a haphazard or random manner. For example, knappers very rarely removed flakes from "top" surfaces in the manner shown on Figure 8b. These flakes would have completely or largely cortical surfaces if they were initial removals across the top of the quarry block and no such flakes occur in the assemblage. Moreover, if the top surface had been previously flaked to produce a slightly convex top surface, the bands would appear curved in plan on derived flakes (as on Fig. 10b). Flakes with curved banding rarely occur in the assemblage. Rather, the banding when present almost always appears on derived flakes as straight lines paralleling one another in plan view or roughly at a right angle to the dorsal-ventral axis in flake profile (Fig. 10a). In sum, almost all flakes were removed at a right angle to the banding planes or the "body" of the flake was always detached from a surface approximating a "side" of the original quarry block (as on Fig. 8c-d, e-f, g-h, i-j, etc.).

A second line of evidence indicating a systematic reduction of the blocks is the consistency of the banding orientation on the derived flakes. Because the bands largely appear as straight lines, their orientation in plan view versus the longitudinal axis of the flake can be measured (Appendix A explains how this measurement was taken). As shown on Figure 11a, measurement of this orientation on the Thedford II unifacial assemblage indicates the bands predominantly run at right angles (0–10° in the method of measurement) in plan to the longitudinal axis of the flake or they parallel it (80–90°).

There is no inherent reason why the banding orientation should strongly follow these trends unless a systematic reduction procedure was followed. This is well illustrated through an examination (courtesy of William A. Fox) of flakes from three quarry reduction stations associated with the chert outcrops. Based on the few diagnostics recovered, these sites are inferred to be largely Archaic in affiliation.

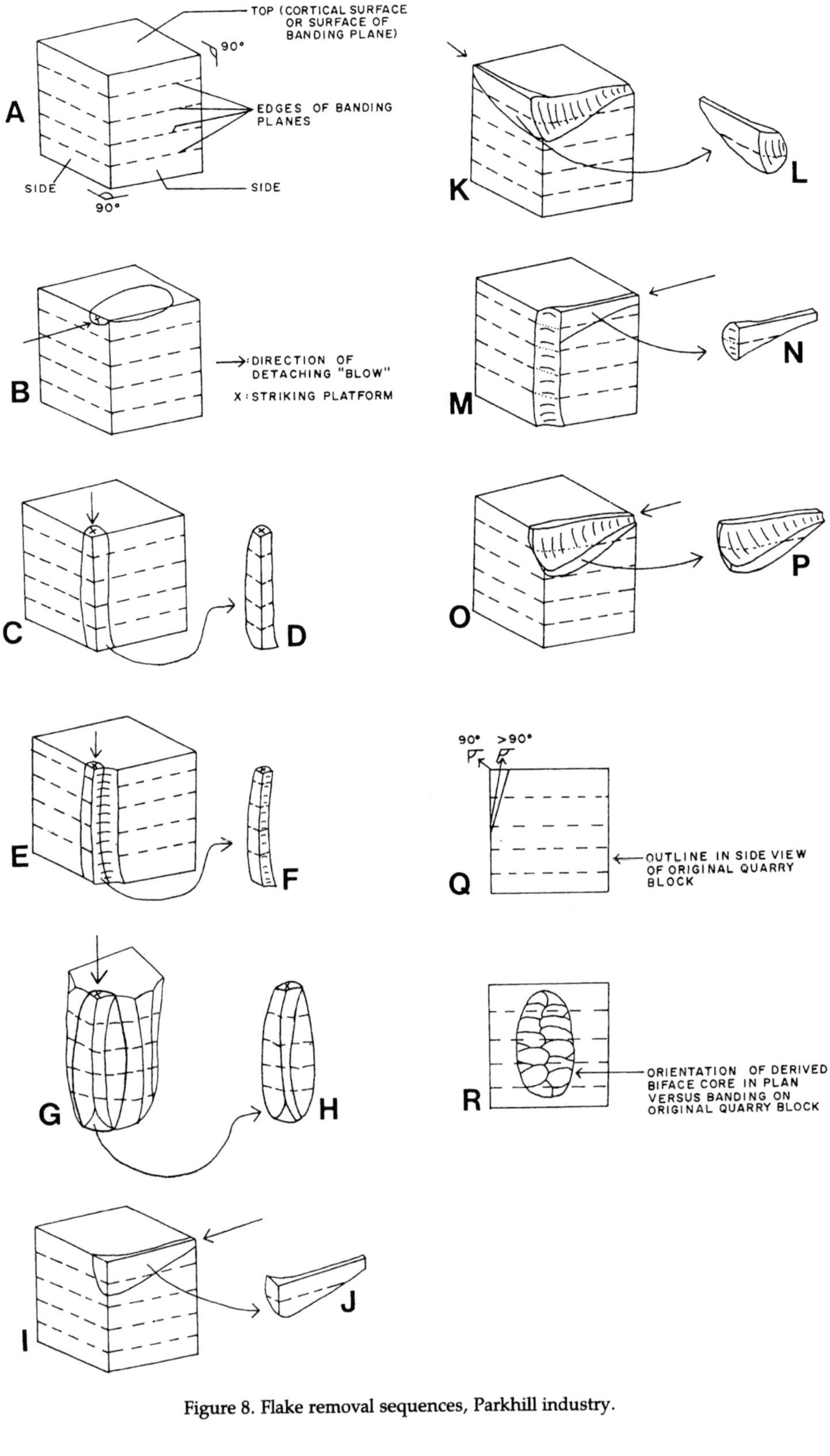

Figure 8. Flake removal sequences, Parkhill industry.

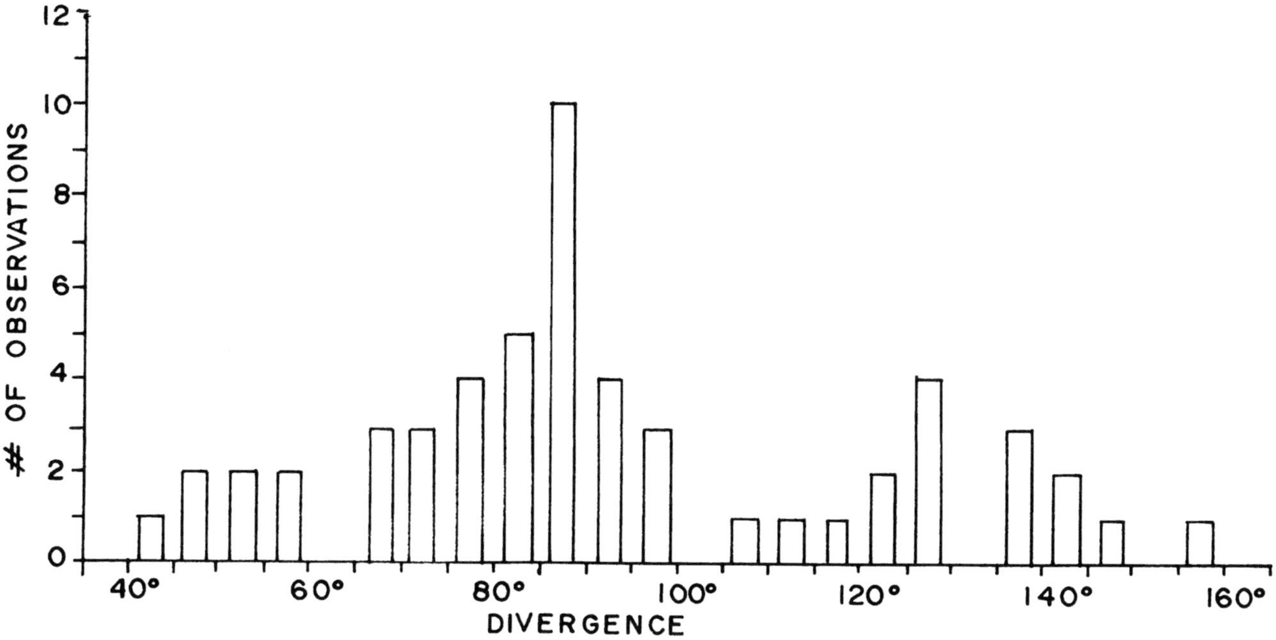

Figure 9. Angles of "side-side" junctures, Collingwood chert bedrock sample.

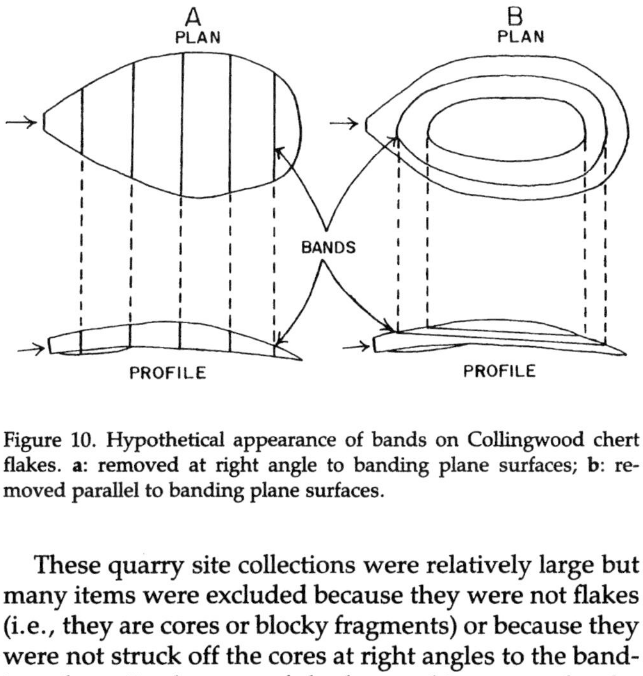

Figure 10. Hypothetical appearance of bands on Collingwood chert flakes. **a**: removed at right angle to banding plane surfaces; **b**: removed parallel to banding plane surfaces.

These quarry site collections were relatively large but many items were excluded because they were not flakes (i.e., they are cores or blocky fragments) or because they were not struck off the cores at right angles to the banding plane. In the case of the latter, this means the derived flakes do not exhibit straight parallel bands in plan view but rather, exhibit curved banding or alternatively, primarily cortical surfaces (that is, they were not removed in the same way as flakes shown in Fig. 8b, 10b). Although not quantified, this form of banding and primarily cortical surfaces appear to be common in these quarry assemblages and this by itself contrasts with the Paleo-Indian materials. As well, it indicates less structure to removals than was employed in the Thedford II assemblage. It also suggests the constraint in the Paleo-Indian assemblages against flake removals across a top surface is not primarily due to characteristics of the raw materials.

Even on the Archaic flakes with straight, parallel banding in plan, the consistent patterning seen in the Paleo-Indian sample is not present. This can be clearly seen when the measurements are plotted for both the Archaic derived and Paleo-Indian flakes (compare Figs. 11a and 12). This denotes an Archaic reduction strategy which is considerably more random than the highly systematic strategy represented in the Paleo-Indian assemblage. It should be noted that Archaic flakes derived from bifaces were excluded from the comparative analyses because biface edges are rarely straight and as a result, we would expect some banding variability to be introduced. However, even with such flakes included in the Thedford II sample, there is still more variation on the Archaic quarry sites.

A final line of evidence indicating systematic reduction of the quarry blocks is the presence of flakes derived from standardized cores such as bifacial ones. These will be described below.

Reduction Sequence I

It has been argued elsewhere (Ellis 1984) that two discrete sequences of standardized quarry block reduction were employed by Paleo-Indian knappers of the Parkhill industry. The first sequence began with a quarry block and throughout reduction, top surfaces were consistently used as platforms for flake removals. Therefore, all blanks produced in this sequence exhibit banding at right angles to the longitudinal axis in plan view and some of the blanks have cortical platforms. All blanks exhibiting one or both of these attributes are referred to as "top-removed" which indicates top surfaces were used as striking platforms.

Three types of blanks were produced in this sequence. Characteristics of these and other blank types to be defined below are summarized in Tables 1–7 and Figures 11 and 13–18. The initial removals from such blocks lack evidence of any previous flaking on their dorsal surfaces and are referred to as *normal top-corner* blanks. These are all flakes which were removed down and along junctures of two side surfaces or corners of the block (Fig. 8c-d). As such, the blanks exhibit a single pronounced, centrally located dorsal ridge separating two unflaked side surfaces or facets thereof. These facets meet each other at right angles (that is, they have "core facet angles" [see Appendix A] of around 95° or less). This results, along with the medial ridge, in a pronounced triangular transverse cross-section. These flakes consistently occur in Parkhill industry assemblages, albeit in low frequencies. Only one occurs in the Thedford II sample but others are reported from the McLeod site (Ellis 1984: Plate 6a) as well as at Parkhill (Deller 1980a: Plate H2). These blanks can be flat in longitudinal section. However, some examples exhibit a pronounced distal curvature where the flake curved as it went through the opposite face of the core.

The second type of flake removed in this sequence is the *secondary top-corner blank*. These are similar to the previous type in terms of cross-section, the presence of a relatively pronounced central ridge, and a generally flat longitudinal section excepting some pronounced distal curvature. They differ mainly in that only one of the dorsal facets is unflaked. The other facet is a scar representing the use of the same platform for a previous removal from the core of a normal or perhaps another secondary top-corner blank. In short, the sequence begins with the removal of flakes as shown in Figure 8c-d and proceeds to the removal of flakes as in Figure 8e-f from near the same area or original corner of the block.

Subsequent to the top-corner removals and the flaking away of the initial unflaked surfaces and abrupt corners of the quarry block, removals were continued from the same platform to produce what are referred to as *top-face* blanks (Fig. 8g-h). These flakes are flat in cross-section with less pronounced dorsal ridges and exhibit low numbers of dorsal scars. The scars are parallel-unidirectional and originate at the platform end of the flake being examined. As a result of the use of tops as platforms, the band edges appear as straight parallel lines in plan which are oriented at right angles to the flake's proximal-distal axis (Fig. 11d).

TABLE 1
Top-Corner Blanks

Variable	Normal Top-Corner	Secondary Top-Corner
Length	—	67.4
Width	27.0	30.5
Thickness	—	11.6
Platform length	—	9.7
Platform width	—	3.8
Weight	—	21.1
Curvature	—	13

Measurements in mm; weight in grams.

TABLE 2
Normal Side-Corner Blanks

Variable	N	R	X	S
Length	3	41.5–62.8	55.43	—
Width	7	21.7–33.2	27.58	3.870
Thickness	7	7.3–11.4	9.79	1.567
Weight	5	7.1–15.2	9.40	3.313
Platform length	4	3.3–7.3	5.86	1.763
Platform width	4	2.0–4.6	3.11	1.106
Curvature	5	5–10	7.40	—

N = number of observations; R = range; X = mean; S = standard deviation.

TABLE 3
Secondary Side-Corner Blanks

Variable	N	R	X	S
Length	5	39.7–64.8	54.46	10.296
Width	10	24.4–46.7	35.03	7.769
Thickness	10	7.2–12.2	9.89	1.762
Weight	6	11.0–27.6	18.93	7.219
Platform length	7	4.3–13.1	8.39	2.763
Platform width	8	1.9–5.5	2.78	1.165
Curvature	6	4–10	6.8	—

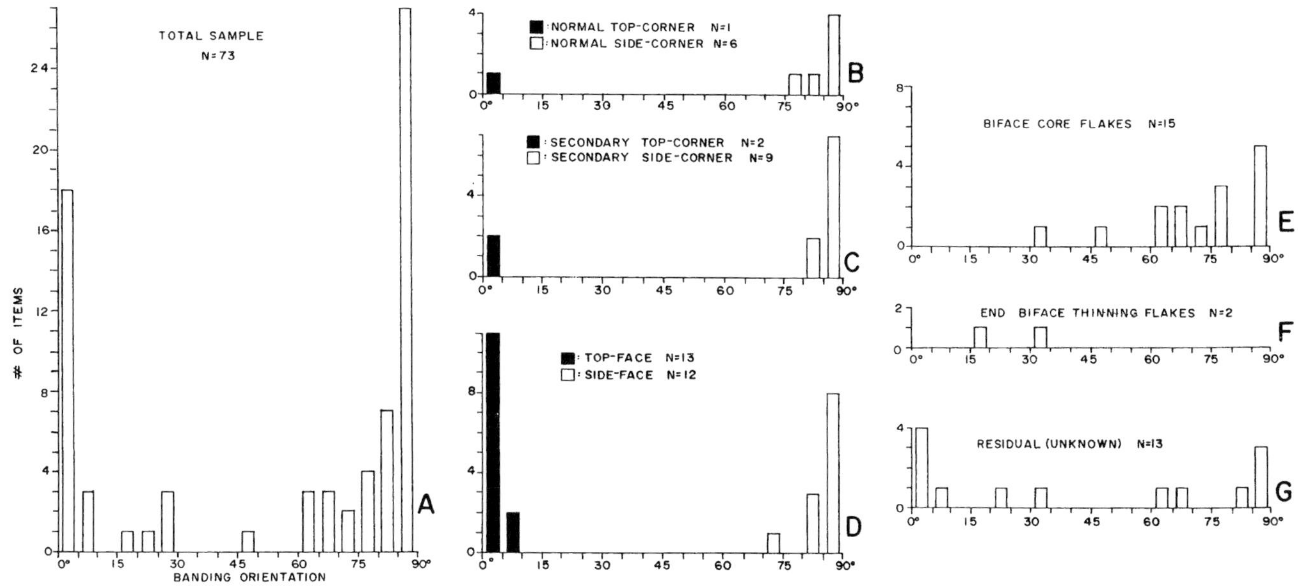

Figure 11. Banding orientation, tool blanks.

TABLE 4
Top-Face Blanks

Variable	N	R	X	S
Length	9	35.5–64.7	50.86	11.793
Width	13	23.7–56.7	33.27	11.088
Thickness	12	3.7–17.3	8.45	3.581
Weight	10	4.8–51.2	16.56	15.053
Platform length	6	4.2–14.7	10.74	3.655
Platform width	5	3.9–6.6	4.79	1.078
Curvature	11	9–15	12.09	—

TABLE 5
Side-Face Blanks

Variable	N	R	X	S
Length	9	24.7–60.1	45.14	12.152
Width	11	16.1–44.2	30.78	7.654
Thickness	12	3.5–12.3	7.79	2.573
Weight	8	5.1–26.7	14.04	7.197
Platform length	6	6.0–18.1	11.88	4.753
Platform width	7	1.6–16.7	6.88	4.931
Curvature	11	5–15	11.7	—

TABLE 6
Flakes From Biface Cores

Variable	N	R	X	S
Length	18	30.2–58.9	41.91	9.315
Width	19	17.4–37.4	30.30	5.847
Thickness	21	4.9–8.6	6.55	1.140
Weight	17	4.0–14.3	7.89	2.917
Platform length	17	5.2–18.3	8.82	3.778
Platform width	18	1.5–4.5	2.68	1.058
Curvature	18	6–12	9.22	—

TABLE 7
Total Number of Blanks Per Type

Type	N
Normal top corner	1
Secondary top corner	2
Normal side corner	9
Secondary side corner	10
Top face	14
Side face	13
Flakes from biface cores	21

The blanks from a large number of Parkhill industry sites including Dixon (Deller 1979: Fig. 24f), McLeod (Ellis 1984) and others, show the consistent presence of large flakes (over 20 g in weight) among the top-face blanks. These large items: (1) exhibit extensive platform preparation; (2) have little or no curvature in longitudinal section indicating, given their size, careful control over removal; (3) are (or were prior to retouch) similar to Old World blades (i.e., twice as long as wide); (4) have dorsal scars indicating the removal of previous

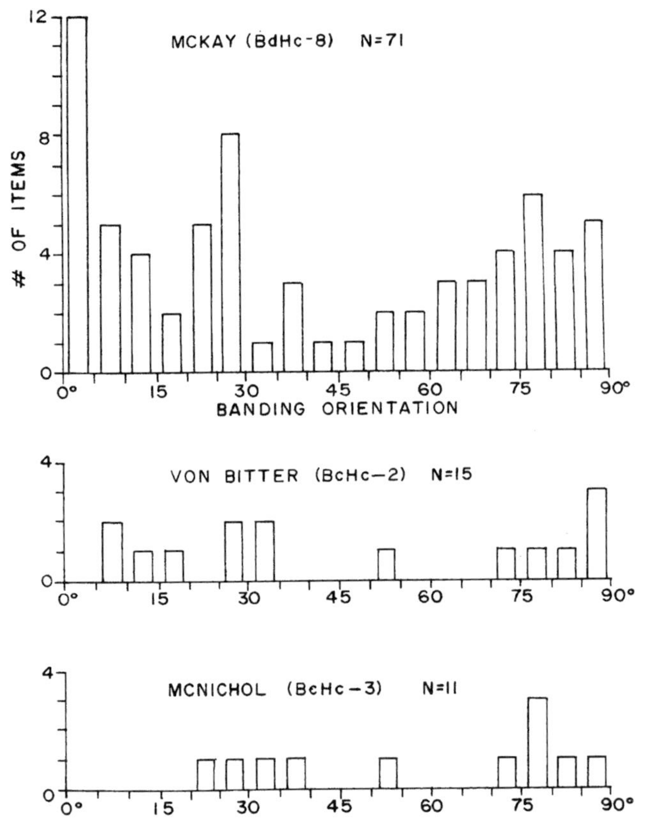

Figure 12. Banding orientation, quarry reduction stations at Collingwood chert outcrops.

flakes from the core of a similar size and configuration; and (5) can show evidence of extensive dorsal preparation to set up the core faces for subsequent flake removals. With regard to this last characteristic, and as an example, Figure 55b (see also Figure 51c) has had two very long (ca. 55 mm), narrow (ca. 15 mm) and thin (ca. 2–3 mm) blades removed down the dorsal surface to create longitudinal ridges to guide subsequent removals. It has been suggested (Ellis 1984:126) that these large top-face blanks indicate the production and use of unidirectional "block" cores of a roughly conical shape like those experimentally used by Callahan (1979: Fig. 1c,d) to produce "blade-flakes." In this interpretation, the smaller top-face blanks would represent trimming of these cores as well as, perhaps, final removals from the largely exhausted core. Significantly, recent work at the Windy City Paleo-Indian lithic workshop in Maine (Payne 1987) has provided, through refitting of flakes, definitive evidence of the use of such conical cores, exactly as predicted at Thedford II based on tool blank morphology (see also Sanders 1983:69–72).

In summary, the first primary sequence of manufacture involved simple gradual reduction of the initial quarry block to a specialized unidirectional block core used primarily to produce large blade-flakes. As will be discussed in subsequent chapters, these blade-flakes were used largely as blanks for fluted points. Throughout this sequence, the same platform (a top) was employed such that the derived flakes all exhibit the banding at right angles to the longitudinal axis in plan view.

Reduction Sequence II

The second reduction sequence began with a quarry block and ended up with a large biface from which flakes suitable for use as tool blanks could be produced. In contrast to the previous sequence, the derived flakes throughout the sequence were struck off almost entirely such that the parallel straight bands on the flakes follow the longitudinal axis. In short, the sides rather than the tops of the quarry blocks served as the platforms for flake removals. Hence, blanks produced in this second sequence are referred to generally as *side-removed* blanks, meaning that a surface approximating a side was used as a striking platform.

One type of blank represents the initial detachments from the quarry block in this sequence: *normal side-corner* blanks. These blanks are struck off along a juncture of a top and side surface of an original quarry block (Fig. 8i-j). Thus, they exhibit a pronounced dorsal ridge separating two unflaked dorsal facets, one of which represents part of an old top of the block and the other, a side. Top facets are almost always cortical (5 of 7) while side facets are always weathered surfaces.

In cross-section, the juncture of the top and side facets on the blanks always has a core facet angle of around 90°. This is to be expected because, as noted above, blocks from outcrops consistently exhibit junctures of this nature. Another notable feature of these flakes is that they were almost always struck off so that most of the flake body was removed from the side rather than the top of the block. As a result: (1) the top facet is narrow in comparison to the side facet; (2) by extension the dorsal ridge is not centrally located but instead is markedly offset towards one lateral edge; (3) the blanks are almost always wedge-shaped in transverse section with the narrow top facet being the back of the wedge; and (4) the band edges appear as straight lines paralleling one another in plan view.

The normal side-corner blanks generally exhibit a pronounced distal curvature where the flake encoun-

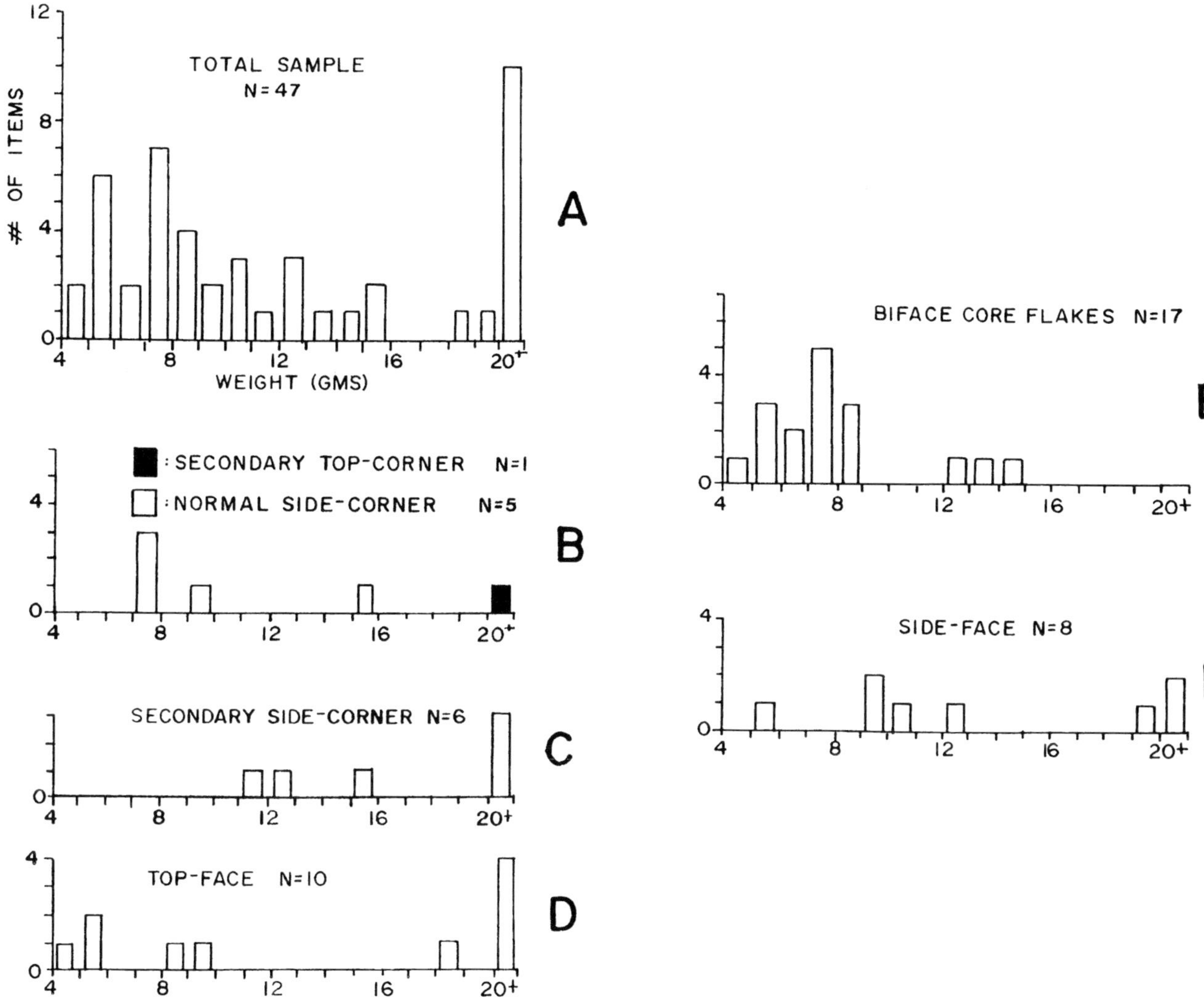

Figure 13. Weight, tool blanks.

tered a side of the block opposite that used for a striking platform. Indeed, these flakes can exhibit at the distal end a facet representing the opposite side or bottom of the core (Fig. 8i-j). Although they lack dorsal scars indicating the previous removals of flakes from the same platform, some exhibit scars at the distal end where there is the pronounced curvature. These previous removals can be from a direction opposite that of the flake under examination and represent previous removals of side-corner blanks from another opposite corner of the block as in the sequence of Figure 8i-j and then Figure 8k-l. Alternatively, and rarely, these scars are transverse to the side-corner removal and employed the old cortical top of the original block as a striking platform. Since these scars originate at the distal curvature of the normal side-corner flake, this suggests these transverse removals were probably normal or secondary top-corner removals. In other words, after a removal as on Figure 8c-d, a removal like that shown on Figure 8m-n was made.

As these blocks were reduced to biface cores, it can be suggested that the previous top-corner removals were designed to thin the initial blocks as it would be difficult to work thick squarish blocks into bifaces. Indeed, it may be that thinner blocks were selected at the quarry for the production of biface cores.

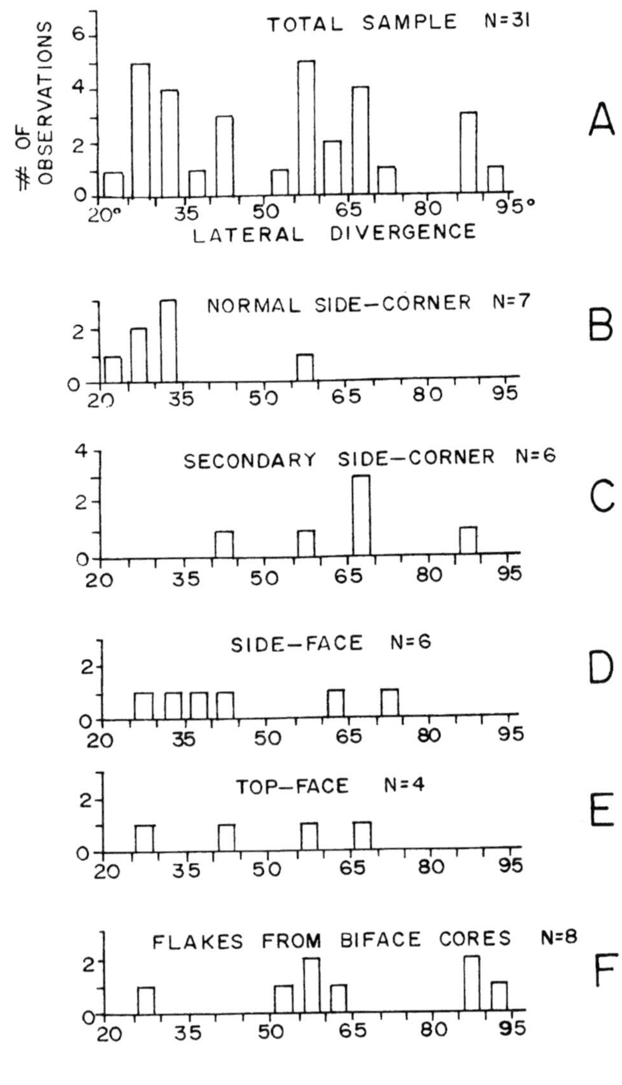

Figure 14. Lateral edge divergence, tool blanks.

In this second reduction sequence, the normal side-corner removals were followed by the detachment of *secondary side-corner* blanks from near the same area of the core (i.e., the knapper proceeded from Fig. 8i-j to Fig. 8o-p). As with the normal side-corner removals, these secondary blanks can have bidirectional scars indicating the presence of at least two opposing platforms on the core. In addition, they tend to have pronounced distal curvature and a pronounced dorsal ridge separating two dorsal facets. The narrow facet is once again a portion of the old top of the core and in the Thedford II assemblage, is always cortical. As these blanks are not initial removals from near a corner, the other facet is generally a completely flaked surface. However, in two cases, the secondary removals expanded down the side away from the corner so that it encompassed all of the scar left by a previous initial removal as well as a weathered surface segment beyond the scar (as on Fig. 8p).

As shown on Figure 8q, if the initial corners of the block were at 90°, the subsequent corner-removed flakes should exhibit core facet angles above 90° and this is indeed the case (Fig. 16b). As a result of this greater divergence, the secondary side-corner blanks have offset triangular rather than wedge-shaped transverse sections. Moreover, probably because the ridge is less pronounced along the corner, the flakes tend to be wider and to exhibit more expanding edges from the platform (i.e., they do not "bite" as deeply during removal).

Presumably, after the side-corner removals, the knapper worked down the side from the corner towards the base of the core flaking away all the original block surfaces not near corners in the process. After these original surfaces and abrupt corners were removed, the continued reduction of the blocks using surfaces with a "side" orientation as striking platforms produced *side-face* blanks. Basically, these are blanks distinguished because they have longitudinal (80–90°) banding, are not from corners and have completely flaked dorsal surfaces. In several cases the dorsal scars are bidirectional suggesting continuity from earlier removals (side-corner blanks) and with subsequent removals from biface cores which, by definition, have two opposing platforms or lateral edges from which flakes can be removed. It is suggested that the side-face blanks are the result of the rough dressing of the core before a point where recognizable flakes from true biface cores were produced. The side-face blanks tend to exhibit more right-angled platforms; have relatively high dorsal flake scar counts as a result of bidirectional working of the cores; can be somewhat large and especially, thick; and either lack longitudinal curvature or have distal curvature.

After the side-face removals, the result was a very large bifacially worked core as has been reported from several Paleo-Indian sites (Gramly 1982:35; MacDonald 1968: 65, 1971:34; Knudson 1973; Frison and Stanford 1982: Fig. 2.92; Wright and Roosa 1966:857). Flakes from these cores in the Thedford II assemblage occur on Bayport and Onondaga as well as Collingwood chert. These flakes should not be confused with the smaller biface thinning flakes described in the waste flake section in Chapter 5 below. *Biface thinning flakes,* as the term is applied herein, are produced from the reduction of small bifaces the size of fluted point preforms. *Biface core flakes* are much larger and result from the reduction

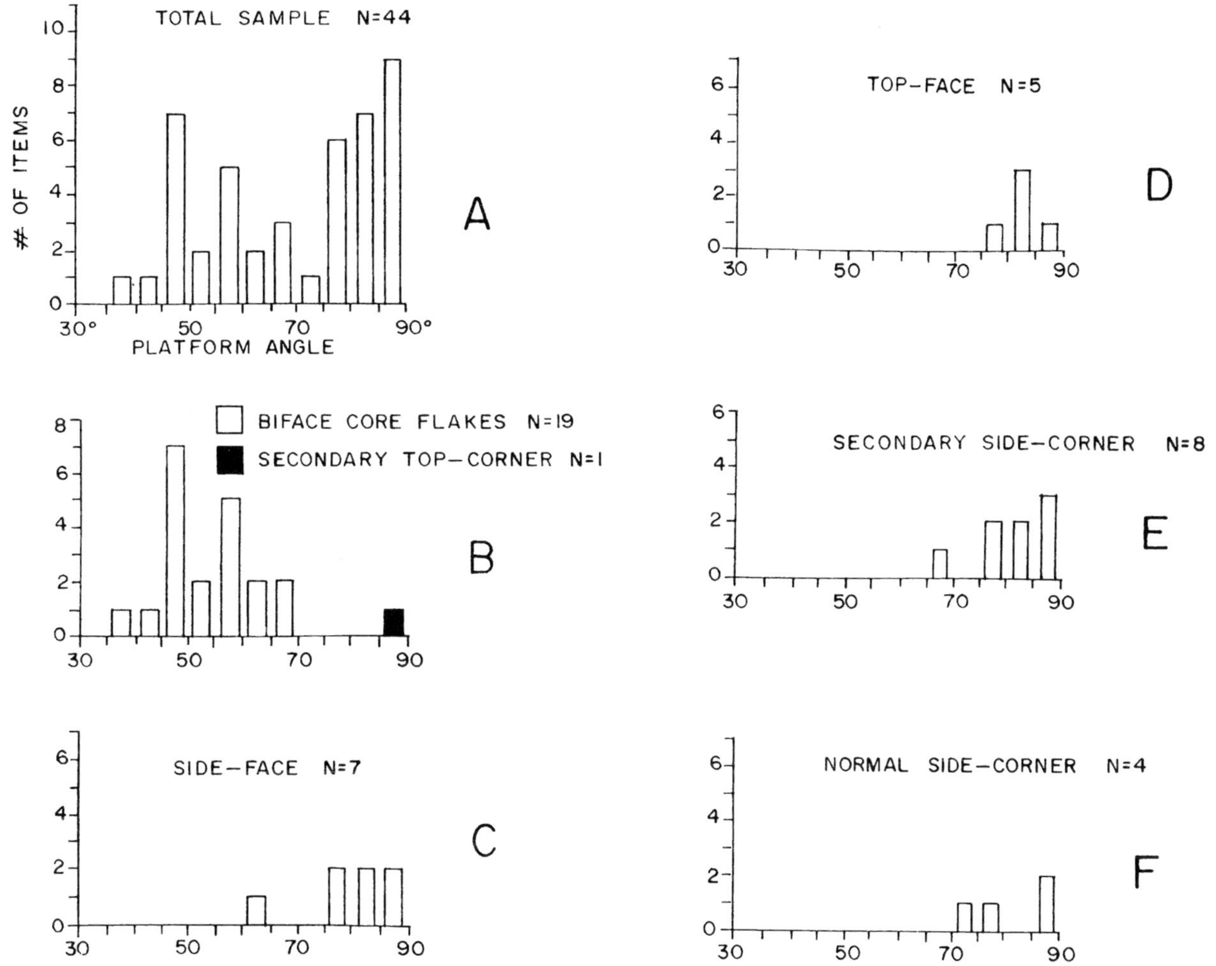

Figure 15. Platform angles, tool blanks.

of very large bifaces. At all sites we have examined, there is no evidence of the actual reduction of such large bifaces. Rather, it seems clear that these bifaces were reduced largely at or near quarries.

In any case, the biface core flakes exhibit several characteristics indicating removal from bifaces, including highly prepared platforms that are usually heavily ground, purposefully faceted and reduced; relatively acute-angled platforms (Fig. 15b); high flake scar counts (although the fact these blanks were often employed for highly resharpened tools such as trianguloid end scrapers means the counts given on Fig. 17g are minimal estimates); often convergent dorsal scars since the platform edges of the biface were not straight but rather, curved in plan; bidirectional dorsal scars; a thin, flat transverse section; generally expanding lateral edges; and some degree of longitudinal curvature.

Another notable feature of these flakes concerns the banding orientation on Collingwood chert flakes. Since the edges of the bifaces are not straight but instead, curved, we would expect banding to vary and this is the case (Fig. 11e). The banding orientation is noteworthy in two other respects. First, because it is largely restricted to 45–90°, this indicates the biface cores were not circular in plan but rather, had recognizeable lateral edges from which most flakes were removed. If a circular core was employed, all angles should be represented. This interpretation is supported also by the fact that a few recognizeable, very large end-struck flakes with transverse dorsal scars are reported from Paleo-

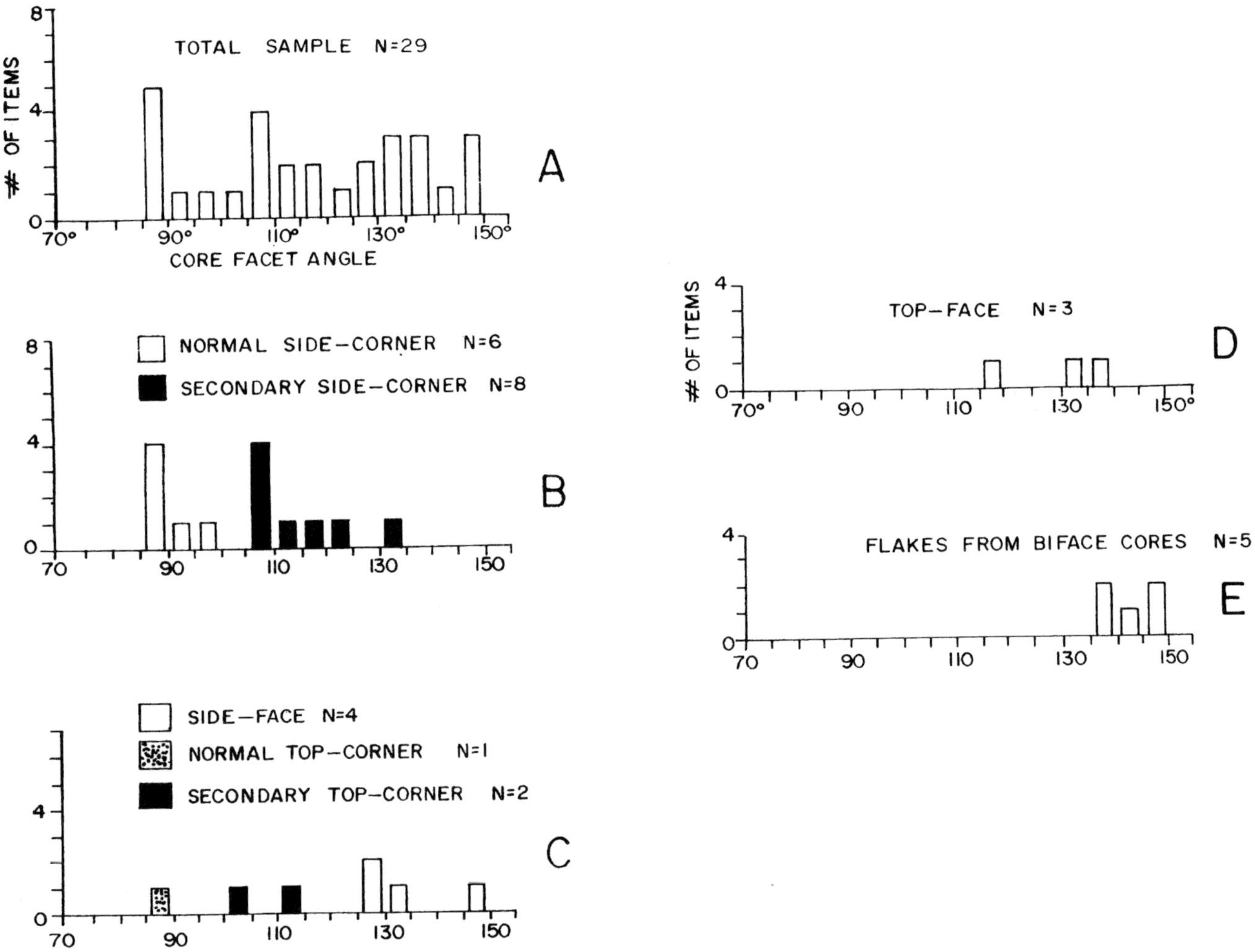

Figure 16. Core facet angles, tool blanks.

Indian sites such as Parkhill (Deller 1979: Fig. 24d) and Crowfield (Ellis 1984:265–66).

Second, the banding orientation on the flakes, because it strongly clusters around or just below 90° (that is, it more closely follows the longitudinal axis of the flake), indicates the biface cores were consistently oriented with respect to the original chert bed or, more properly, quarry block selected from the bed section. In particular, they were oriented such that the lateral edges and faces of the core faced what would have been the sides of the original quarry block while the ends of the cores were placed towards the original block's top surfaces (as on Fig. 8r). Phrased another way, the biface cores were oriented such that the banding would appear on their surfaces at right angles to their longitudinal axis. We should expect different banding orientations if the cores were oriented in any other way relative to the original quarry block.

The known end removals from biface cores in Parkhill industry assemblages are consistent with this interpretation. Since these blanks were removed from the end of the biface core, we would expect them to have transverse (0–5°) banding. On the two known examples from the Parkhill site, this is the case. The orientation of the lateral edges of the biface cores towards side surfaces of the original quarry block is suggestive of continuity in removals from the side-face blanks to the blanks from biface cores. In short, we would expect most of the removals detached in dressing the core to form a biface to be struck off from the same direction

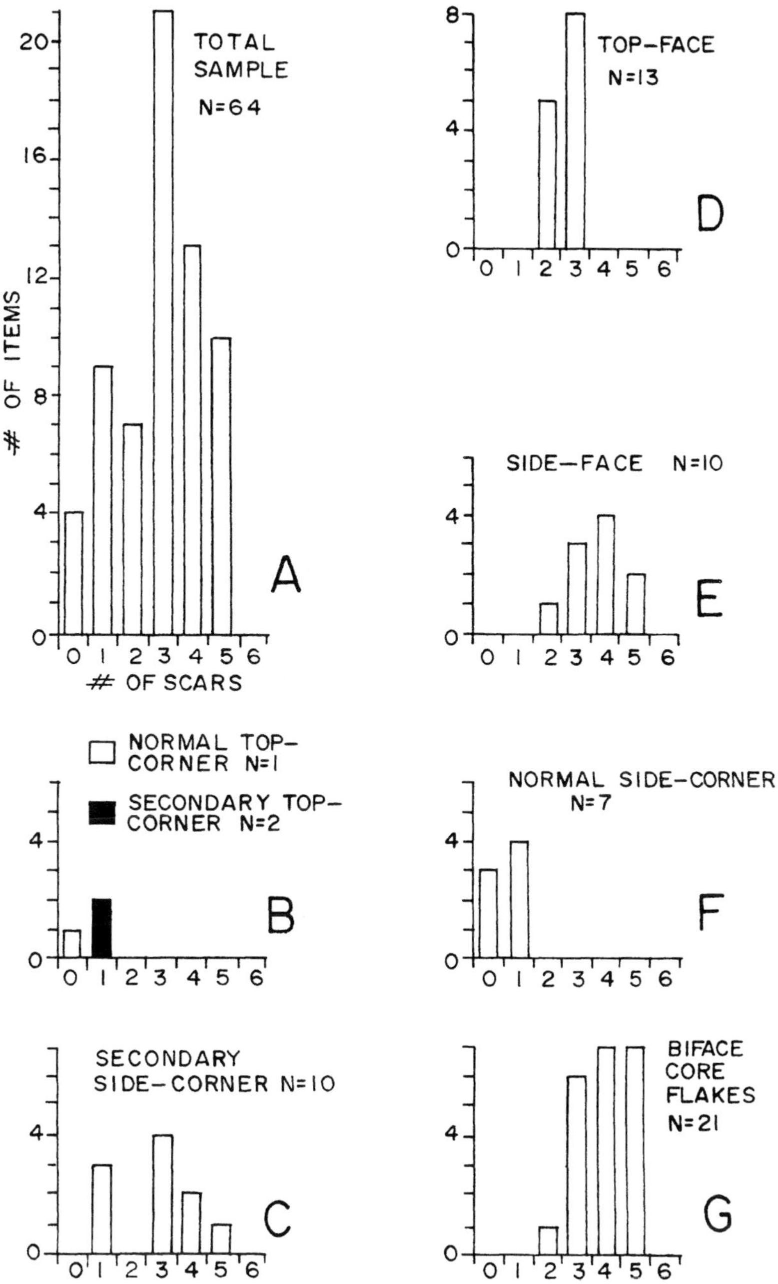

Figure 17. Number of dorsal scars, tool blanks.

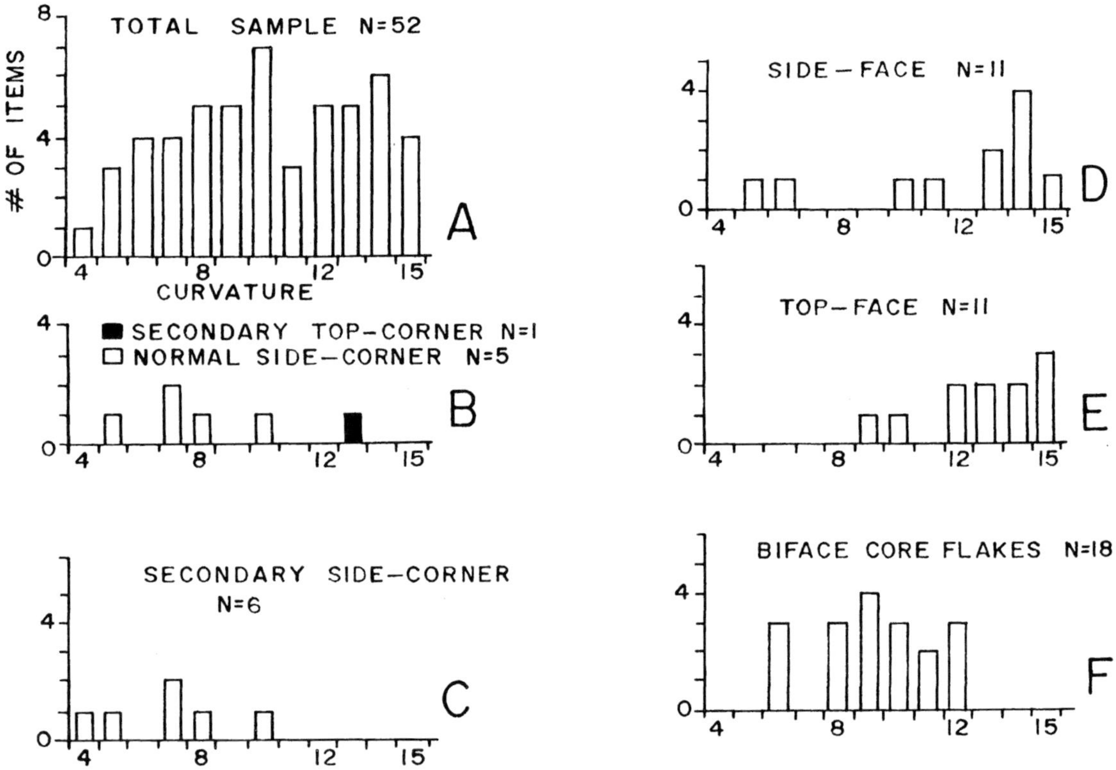

Figure 18. Curvature, tool blanks.

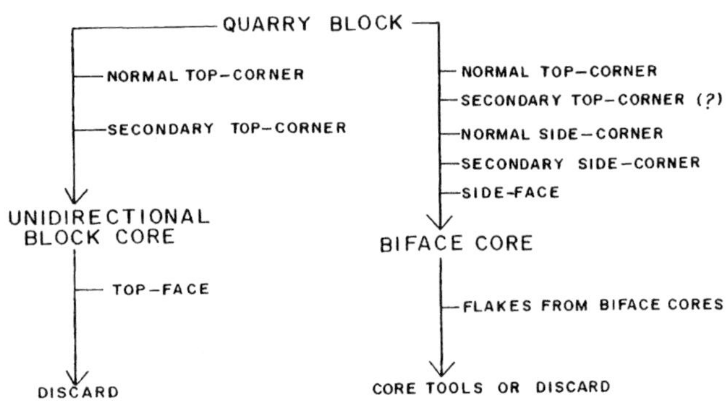

Figure 19. Sequence of flake removals, Parkhill industry.

as where the lateral edges of the biface core would be developed—the sides of the original quarry blocks.

In summary, the second suggested sequence of block reduction began largely with side-corner removals, proceeded through a sequence of side-face removals and ended with a large bifacially worked core from which flakes suitable as tool blanks were detached. This is essentially the same sequence of core and blank production as was suggested for the Barnes site assemblage by Wright and Roosa (1966). This sequence of removals and the previous one are illustrated in Figure 19.

CHAPTER 3

Bifacial Tools/Preforms

In this and the following chapter, we describe and discuss the artifact assemblage recovered from the Thedford II site. The tools are described by general category (biface, uniface) and class (fluted bifaces, other biface tools, end scrapers, side scrapers, combination notch/borers/denticulates, bend-break tools, other unifaces). Where applicable, types that can be recognized within the more general tool classes were isolated.

These descriptions and discussions are detailed for several reasons. First, we believe that comparative studies between sites often have been limited, or precluded by the paucity of published data. Although we recognize that it is impossible to consider every potentially informative characteristic, we hope to provide a more comprehensive basis for future comparative analyses.

Second, we attempt to document fully patterned variation in the assemblage. It has been suggested elsewhere (Ellis 1984:4–5; Moeller 1980:82; Meltzer 1984) that much of the past work on Paleo-Indian materials has tended to emphasize their homogeneity or similarities in order to place the materials into time/space frameworks or to simplify description. As such, variability has been largely ignored. Binford (1972:199) observed that this was a general characteristic of much past work on archaeological materials. However, in the Paleo-Indian case, homogeneity has been emphasized to an even greater degree and over vast spatial areas, largely to support models of a very rapid spread of fluted point makers with a similar lifeway throughout the continent. In sum, in Paleo-Indian studies there has been a reluctance to abandon a normative framework or to recognize that elements of this framework still form the basis for much of modern thinking. Note that this is not a denial of the fact that Paleo-Indian materials may be more homogeneous than later materials. It simply means that there has been less concern with variability. Neither do we believe that the demonstration of variability in these materials necessarily leads to the rejection of the "rapid spread" model as some investigators have implied.

Third, and this is an extension of the perceived emphasis on the homogeneity of Paleo-Indian materials, the types described are not simple morphological ones isolated to describe or to place materials in time/space frameworks. Rather, they are isolated because they represent the many direct or proximate causes of variability which are operative in any system of lithic tool production and use. These include: raw material characteristics, primary and secondary manufacturing procedures, recycling, resharpening, discard in manufacture and presence or absence of hafting or variability in hafting methods. As well, the types are thought to signal use differences within the assemblage even though they are not defined on the basis of use-wear studies. As argued in detail elsewhere (Ellis and Deller 1988), use significance is implied by the highly consistent and complex set of decisions in shaping and manufacture which are homogeneous within items assigned to a type and contrast markedly between types (for example, fluted points versus gravers or micro-piercers) and the fact that the variation between types cannot be explained as a product of raw material, idiosyncratic behaviours, resharpening or other sources of variability mentioned above. Because the types are significant in terms of all or several of these factors, more detail is necessary in order to make explicit the characteristics related to each. Therefore, a more complete understanding of morphological variability at this proximate level should be achieved.

Finally, despite the above statements, cultural-historical frameworks are obviously important. Some of the implement types described are distinctive enough that they can serve as diagnostic indicators of Paleo-Indian materials. Furthermore, as will be discussed in later chapters, they may be indicative of regional or temporal variation in Paleo-Indian tool kits. Thus, they are worthy of more detailed descriptions.

Fluted Bifaces

The Thedford II bifaces related to fluted point production and use are described below. After briefly outlining the nature of the sample, information bearing on fluted point manufacture is presented. Finally, comparisons of the fluted bifaces to other collections from the Great Lakes area are given in order to demonstrate that the Thedford II examples are Barnes points of the Parkhill complex. The distribution of the points and other tools by the lithic material types is given in Table 8.

The Sample

Fluted Bifaces

After refitting, 32 fluted bifaces are represented. In addition, three unfluted bifaces which are probably preforms for fluted points are included in the collections. The characteristics of the fluted points and preforms are summarized in Tables 9 and 10. As well, characteristics of each relatively complete fluted point or preform are given in Appendix B.

The sample includes 12 finished points. Four are almost complete except for recent breaks (Figs. 20a-d, 21a-b). All but one (Fig.20d) lack evidence of resharpening and are essentially in mint condition. They appear to be part of a tool cache in A-Northeast (see Chapter 6) along with some preforms to be described below. One other item (Fig. 20e) is also relatively complete but it has been heavily resharpened and exhibits a foresection impact scar and a broken ear from use as a projectile tip.

Other points are represented by one tip end, two ears and four bases (Fig. 20f-j). One ear and two bases probably resulted from breakage in projectile use but the others exhibit recent breaks. Presumably, the rest of these latter items are still at the site in peripheral areas.

Twelve fluted preforms are represented. All are identified as preforms by the presence of one or more of the following characteristics: (1) extant evidence of facial, basal or tip preparation for fluting; (2) evidence of breakage in manufacture; and (3) a lack of lateral basal grinding. Seven of these bifaces are relatively complete (Figs. 22a-e, g, i, 23a-b) and at least four (Fig. 22b-d, g) are associated with the probable cache. Other preforms are represented by a base produced by a recent break (Fig. 22f) and four tip ends (Fig. 22h,j; 24a-b). Three of the latter were produced by old, presumably manufacture, breaks. One of these (Fig. 22j) has been recycled. The junctures of the lateral edges with a transverse snap have been unifacially retouched such that these junctures appear rounded in plan and the break has been narrowed. A preform reworked in a similar manner was

TABLE 8
Distribution of Tools/Preforms by Raw Material

Type or Class	N	C	B	O
Fluted point	12	6	6	0
Fluted preform	12	10	2	0
Other fluted bifaces	8	6	2	0
Unfluted preforms	3	3	0	0
Concave-based bifaces	2	2	0	0
Alternately beveled bifaces	5	1	4	0
Small oval bifaces	2	2	0	0
Other bifaces	3	3	0	0
Triangular end scrapers	10	6	3	1
Offset end scrapers	6	6	0	0
"Fluted" end scrapers	2	2	0	0
Narrow end scrapers	5	5	0	0
Combination narrow/wide end scrapers	3	2	1	0
Proximal end and side scrapers	8	7	1	0
End and concave side scrapers	2	2	0	0
End and convex side scrapers	2	1	1	0
Other end scrapers	9	8	1	0
Side scrapers	15	14	0	1
Combination notch/borers/denticulates	7	7	0	0
Bend-break tools	3	3	0	0
Backed and snapped unifaces	6	6	0	0
Piercers	18	17	0	1
Denticulates	3	3	0	0
Retouched flakes	6	6	0	0
Other unifacial tools	7	7	0	0
Totals	159	135	21	3

C = Collingwood; B = Bayport; O = Other.

TABLE 9
Fluted Point Variables

Variable	N	R	X	S	C.V.
Length	3	88–105	94.5	—	—
Width	5	22–27.2	25.5	2.158	8.46
Thickness	6	5.2–7.5	6.1	0.945	15.40
Basal concavity depth	5	3.0–5.0	3.6	0.820	22.78
Basal width	4	16.5–19.2	18.4	1.350	7.34
Flute width	15	5.4–17.1	12.2	3.591	19.51
Flute length	9	19.0–82.5	54.1	21.854	40.37
Lateral grinding length	9	13.5–20.6	16.9	2.556	15.10

N = number of observations; R = range; X = mean; S = standard deviation; C.V. = coefficient of variation.

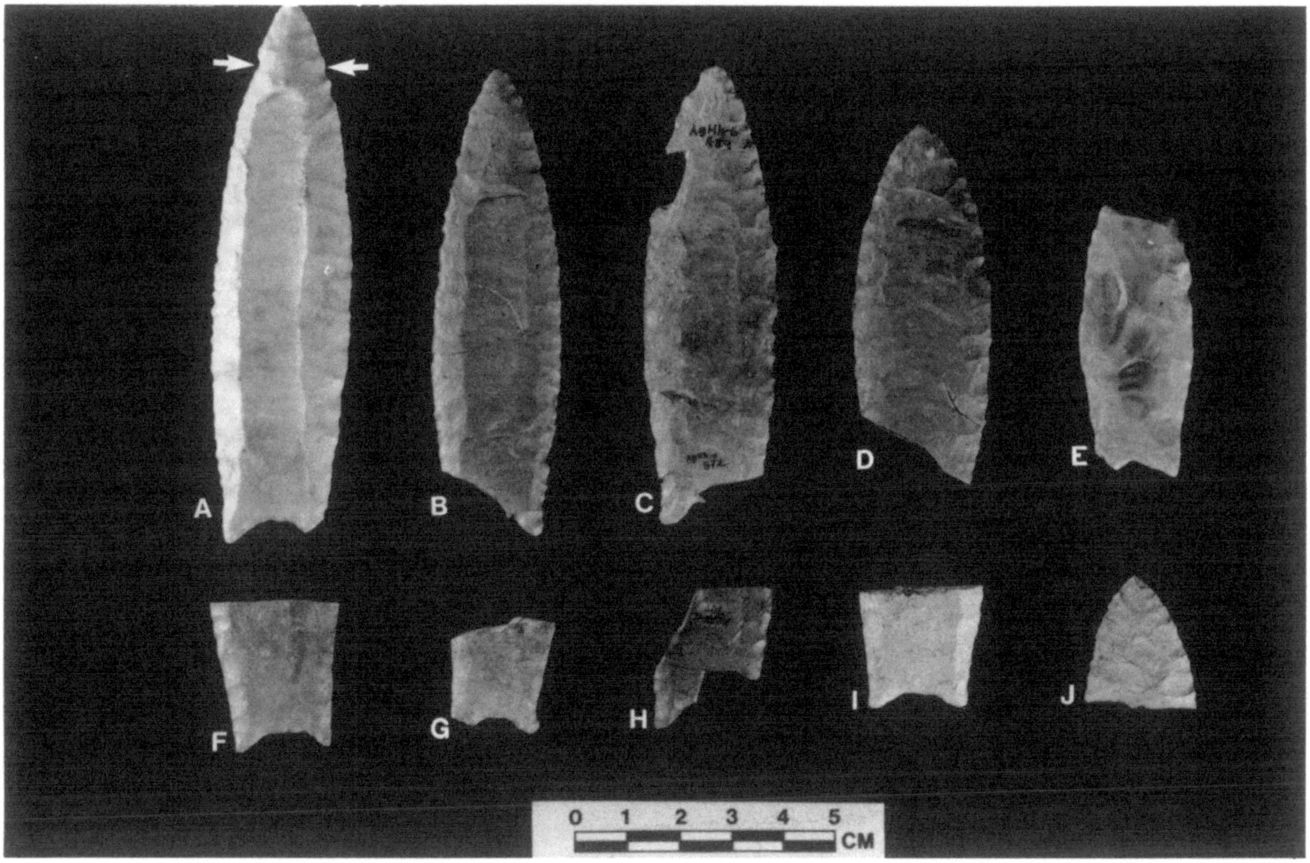

Figure 20. Fluted points, Thedford II site. Arrows show possible tip narrowing remnant.

TABLE 10
Fluted Preform Variables

Variable	N	R	X	S	C.V.
Length	5	32.1–92.0	58.6	23.600	40.27
Width	7	16.7–28.6	24.1	3.767	15.63
Thickness	8	3.6–8.3	5.9	1.356	22.98
Basal concavity depth	7	1.9–6.1	3.9	1.421	36.44
Basal width	4	16.1–19.4	18.1	1.407	7.77
Flute width	20	5.6–17.0	11.9	3.191	26.90
Flute length	8	22.1–45.3	34.8	8.751	25.14

reported from the Barnes site (Wright and Roosa 1966: Fig. 4c). At least one of the preform tips exhibits a platform remnant of the original flake on which it was made.

The remaining fluted bifaces are too incomplete to be assigned to points or preforms. These eight items include: three midsections; two lateral edge fragments and three tips (Fig. 24c-h). Two of the tips were produced by recent breakage. On one, it has been possible to replace a channel flake segment (Fig. 24c). This indicates the point was fluted on the site but, for reasons to be outlined later, does not indicate that it is a preform.

The above sample can be characterized as follows. These bifaces can be very long, but short preforms suggest smaller points were also produced (compare Figs. 20a and 22c-d). Among the largely complete unresharpened points and preforms, there is a suggestion that length is governed by the raw material employed. The longer artifacts are exclusively made of either Bayport chert (e.g. Fig. 20b-d) or the soft, mottled, variety of Collingwood chert (e.g. Figs. 20a; 22b) while shorter items (e.g. 22c, d) are of the common, hard, banded Collingwood. The points are generally narrow and thick with width to thickness ratios of 3–4.5:1. Lateral basal edges moderately expand from at or just above the narrow (under 20 mm) base to a maximum width around, or if the point is unresharpened, just below midpoint.

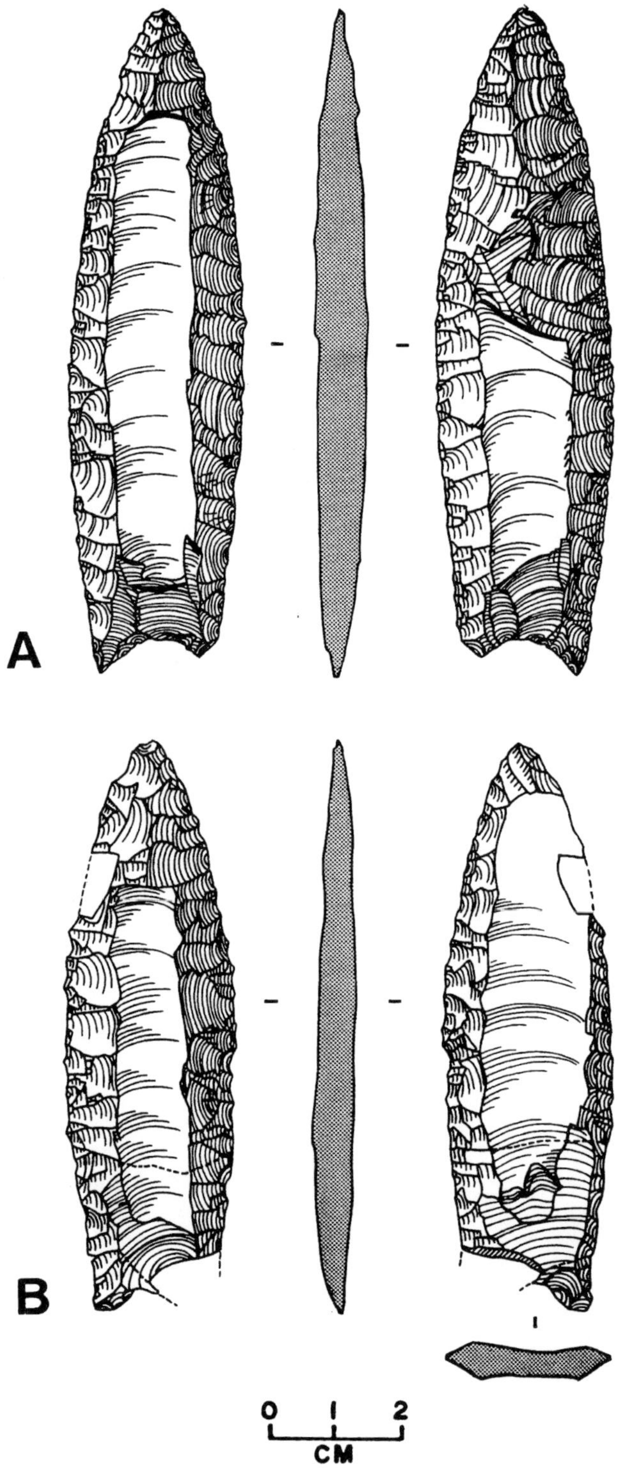

Figure 21. Two fluted points from A-northeast. **a**: Collingwood chert; **b**: Bayport chert.

Most finished points (7 of 9) have a constriction or waist just above the base. Hence, they often have fishtails which in some cases (e.g. Fig. 20g) are quite pronounced. Fishtails also occur on five of the preforms. All finished basal ends have light lateral grinding which extends for only a short distance (under 20 mm) from the base. All also exhibit light grinding in the basal concavity.

The points are very well fluted with a tendency for parallel-sided flutes extending for three-quarters or more of point length (ca. 90% of available sample) on at least one face. On 10 of 15 specimens, the flute extends to or within 15 mm of the tip end on at least one face. No examples have more than two flutes per face and in fact, double fluting is rare. Two flutes are found on one face of three of the nine finished points and on one preform. Flutes tend to be well-centered but two points and one preform have relatively off-center flutes to the left edge on one face (i.e. Fig. 22e).

The bases of the finished points (9 of 10) have been completed by the removal of a broad (10–15 mm), short (10–20 mm) flake over the primary flute scar(s) on both faces. This technique, referred to by Roosa (1965:97) as "Barnes finishing," also occurs on at least one face of six preforms. These removals served to deepen the base of the flute. Frequently, they obscure basal preparation for fluting. The distal end of these finishing flakes often feather out or "blend" with the main flute scar which makes them difficult to discern. However, in several cases (e.g., Fig. 21), there is a hinge or step where the thinning flake terminated. The one point which lacks this Barnes technique has small, short, steeply beveled or abrupt flake removals across the basal concavity on both faces in a manner similar to Folsom (Fig. 20F). We suspect this basal finish is a result of the re-chipping of a new basal concavity onto a point previously damaged in use.

A final characteristic worth noting is a marked tendency for the banding orientation on Collingwood chert points to be at right angles to the longitudinal axis of the point (0–5° in the method of measurement: see Fig. 25). This indicates a great deal of consistency in blank selection for points as will be outlined later.

Unfluted Preforms

Three unfluted bifaces in the assemblage are believed to have been intended as preforms for fluted points (Table 11; Figs. 23c-d, 26a-c). All three have "points of thickness" on their surfaces left by hinge or step-terminated flake removals. Attempts were made to correct these errors by detaching flakes from the opposite mar-

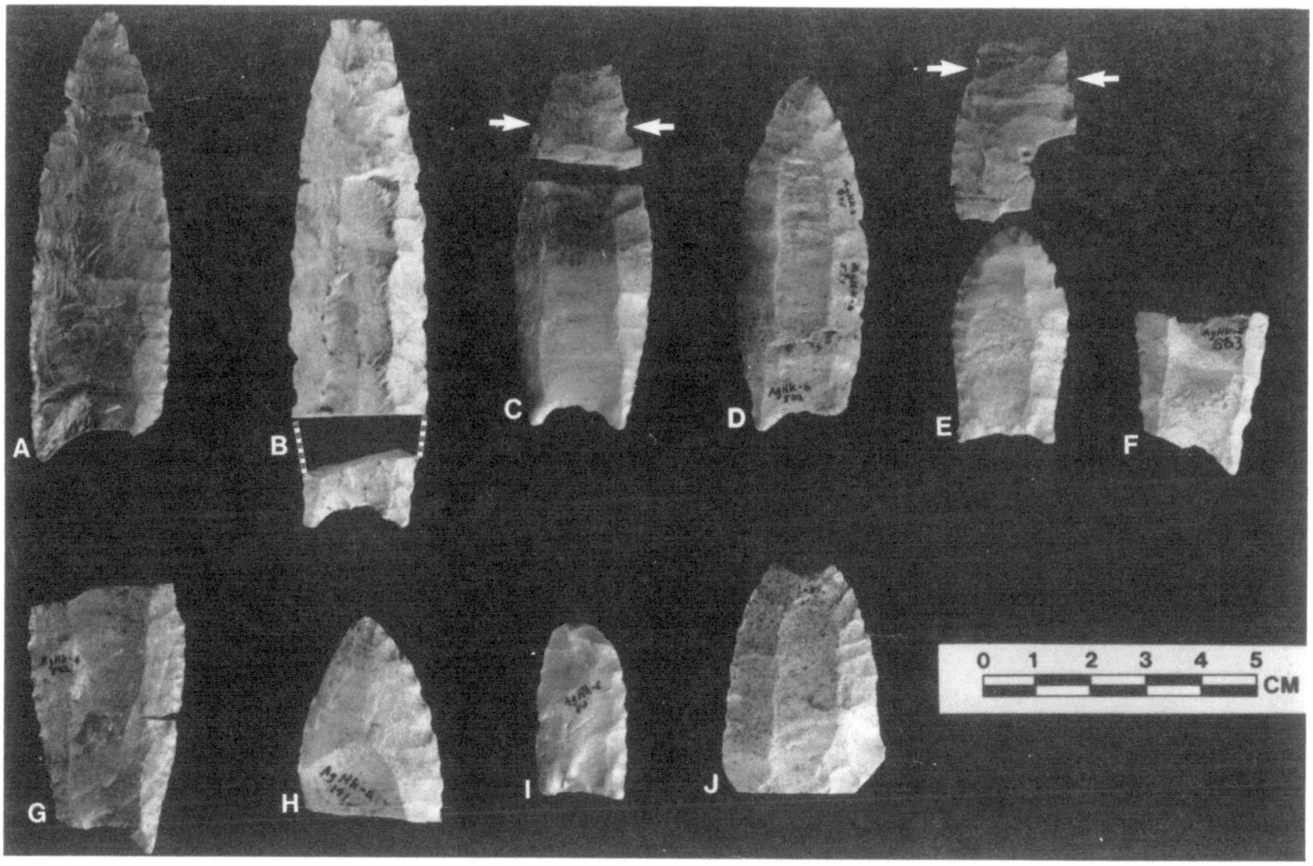

Figure 22. Fluted preforms, Thedford II site. Arrows show deliberate tip narrowing.

TABLE 11
Unfluted Preforms

FC#	Length	Width	Thickness
36	58.1	30.4	10.9
35	67.7	30.1	10.0
592	75.5	30.0	12.3

FC# = field catalog number.

gin or adjacent points on the same margin but these were unsuccessful.

Two of these bifaces (Fig. 26a-b) exhibit 13 or more large, broad, expanding, laterally struck thinning flake scars on their faces which traveled from half to two-thirds of the way across the biface surface. One also has an end thinning flake scar on one face. In size and shape, the lateral and end scars resemble the lateral and end biface thinning flakes recovered in the waste flake collections (see Chapter 5).

These bifaces have material banding at right angles to the longitudinal axis (0–5° and 5–10°; see Fig. 25). The artifacts are oval in plan with slightly convex lateral edges. Maximum width occurs around midpoint and tip ends are somewhat rounded or blunted. The lateral edges appear almost serrated in plan. The tips of the serrations are remnants of continuously ground and beveled platforms created to detach the last series of thinning flake removals. In longitudinal section, the bifaces are straight with no curvature and have sinuous edges. The transverse section is biconvex. One biface has a remnant of the platform of the flake upon which it was made at one end.

The final biface (Fig. 26c) is straight in profile and has a marked biconvex section in end view. As with the previous bifaces, it exhibits banding at right angles to its longitudinal axis. However, it differs from the previous bifaces in that: (1) the edges are less sinuous in plan and profile; (2) the tip end is less blunt; (3) the lateral edges are more convex with maximum width just below mid-point; and (4) flaking consists of smaller, more parallel-

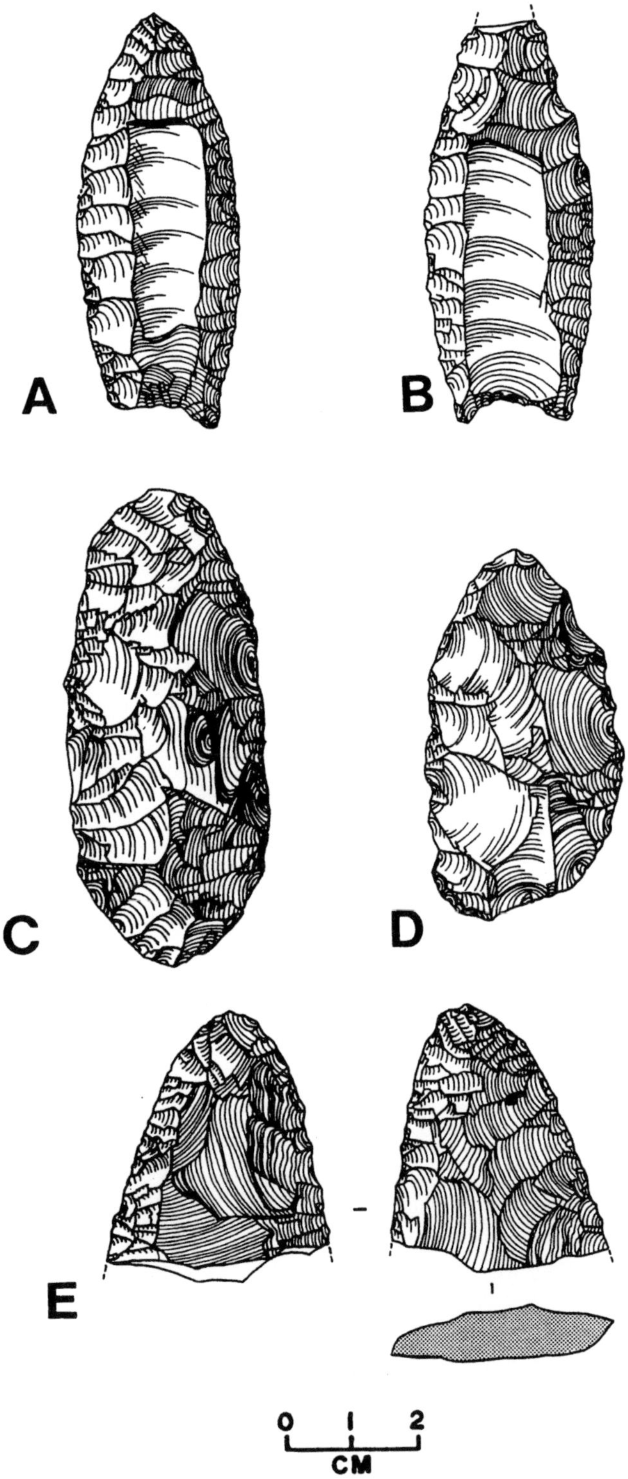

Figure 23. Preforms and biface tool. **a-b**: fluted preforms; **c-d**: unfluted preforms; **e**: tip of alternately beveled biface tool.

sided to only slightly expanding scars which extend only to the midline. An exception to the smaller scars is a broad scar removed in an attempt to correct a step-terminated removal. Two small end flakes were removed from one face of this biface.

We suggest these bifaces are preforms for several reasons. Obviously they are unfinished as they retain parts of prepared platforms and exhibit errors in flake detachments which provide a rationale for discard. Moreover, given that they are unfinished, they lack specialized features which would allow manufacture into other biface tool types known for this industry. Finally, they are of a suitable size to be made into points, exhibit right-angled chert banding which is characteristic of the fluted bifaces, and at least the third item approximates the points in outline shape.

Fluted Point Manufacture

Information on fluted biface manufacture from the site is synthesized here. Detailed lithic reduction models for such items have been presented in archaeological literature (e.g. Callahan 1979). These models have limited applications in the Thedford II analysis because the earlier stages of point manufacture are poorly represented. Also, these models may be applicable only to certain specific fluted point industries. Instead of employing specific models, we use a more simple or general model of biface production derived from experimental literature on biface manufacture (Bradley 1974:192; Callahan 1979:33–37; Henry et al. 1976:58–59; Muto 1971; Newcomer 1971:85–91).

It is clear that the production of bifaces requires that attention be given to: (1) the creation or availability of edges (striking platforms) suitable to allow thinning of the tool (margin production); (2) production of a suitable thickness and cross-section (thinning); and (3) creation of a suitable plan outline and surface finish (finishing). Where bifaces are made from large flake blanks or core blanks, these three goals are largely achieved sequentially by gradually reducing the original blank through a series of progressively more refined biface forms. Knudson (1973:145) calls this serial biface reduction. Thus, given that large blanks were employed here, these three knapping goals can be roughly equated with three stages of biface manufacture.

Most of the points from Thedford II probably were made on flake blanks rather than core blanks. The platform remnants at the ends of some preforms indicate flake blank use as do similar characteristics on Barnes bifaces at other sites (Ellis 1979:38, 128; Storck 1983). Perhaps though, some of the larger points such as Fig-

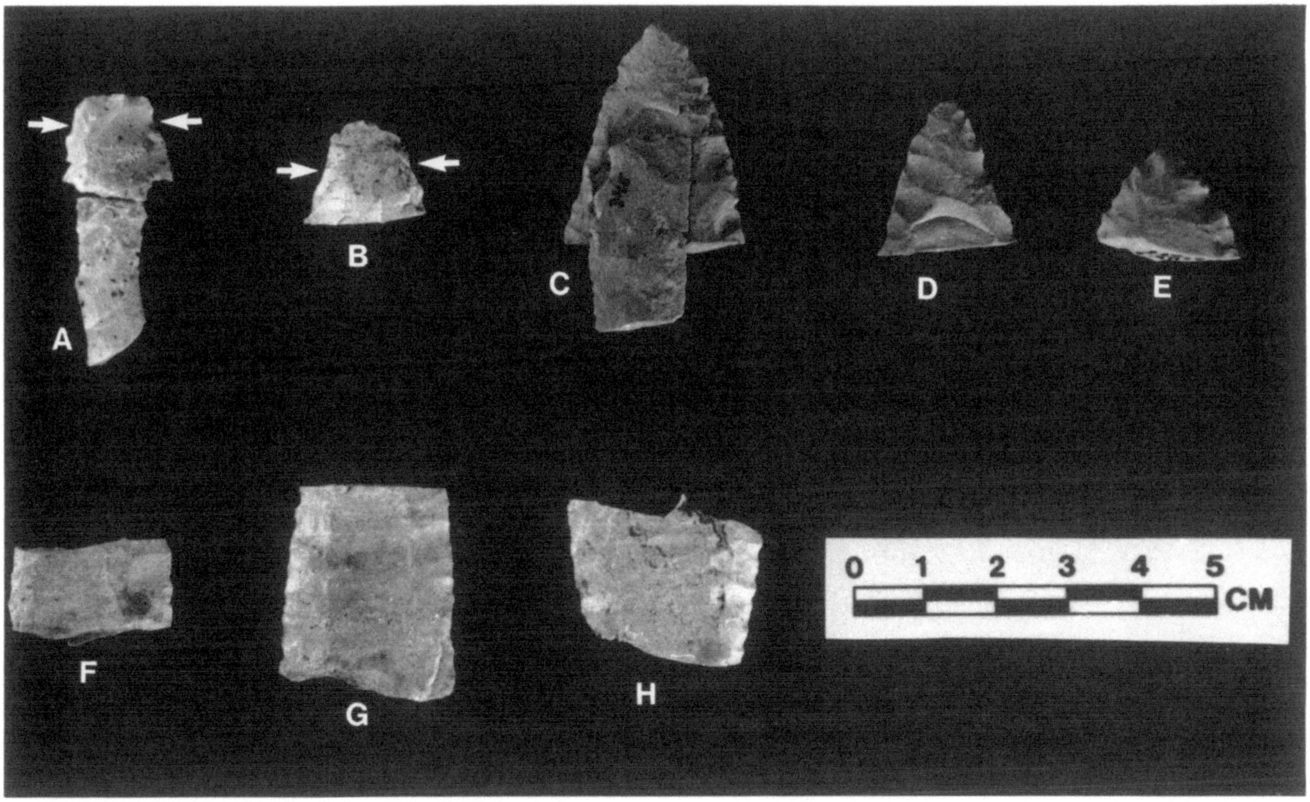

Figure 24. Fluted preforms (**a-b**) and fluted bifaces (**c-h**). Arrows show location of deliberate tip narrowing.

ure 20a were made on core blanks. Based on a combination of features, it appears that the majority of the points were made on blanks of the "top face" type described earlier. First, the top-face blanks have banding at right angles to the longitudinal axis, which is consistent with the banding on the vast majority of the fluted bifaces. This assumes the preforms were largely oriented with the intended tip or base of the point towards the platform end of the original flake. Platform remnants at the ends of the bifaces support this contention. Second, the fact that points were clearly made on large flakes is consistent with the use of top-face blanks since these blanks are consistently the largest available on Parkhill industry sites. Third, the large top-face flakes are the only ones which could be reasonably regarded as "blade-flakes" which, because of their outline shape, are eminently suitable as blanks for points (Callahan 1979:66). Finally, the largest top-face blanks exhibit little curvature. Flat blanks such as these are much easier to work into bifaces (Bordes and Crabtree 1969:9). Moreover, flat flakes facilitate the production and maintenance of a very straight profile on the biface. This is necessary for ensuring the success of extensive fluting in later stages of manufacture (Judge 1973:166).

The initial alteration of the flake blank, the margin production stage, is not represented in the Thedford II collections based on the criteria for recognition outlined by Callahan (1979:67). The subsequent thinning stage is represented by two unfluted preforms (Fig. 26a, b). These are relatively regular in outline and cross-section shape. Also, they exhibit broad flake scars originating at the lateral edges which removed a considerable surface area of the preform and the removals from opposite margins markedly undercut each other at their distal ends. Clearly this indicates a concern with thinning and the achievement of a suitable cross-section.

The remaining preforms are attributed to the finishing stage. One (Fig. 26c) represents a pre-fluting step. This biface has the most regular outline of the unfluted preforms, approximates the finished points in plan shape and instead of broad, thinning flake scars, has smaller, narrower scars similar to those on the finished points which removed little in the way of the artifact's surface. In addition, the lateral removals from each edge

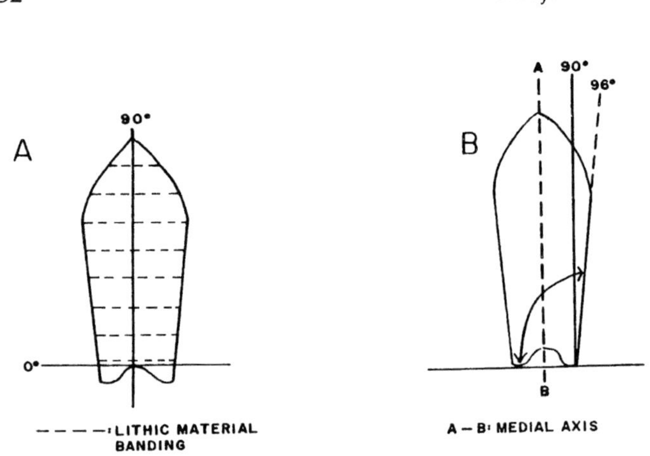

Figure 25. Measurement of banding orientation and face angle, fluted bifaces.

do not markedly undercut each other. These characteristics suggest that thinning is not a major concern.

The next step in the finishing stage involved the preparation of the biface for fluting. All of the fluted preforms in the present assemblage which are complete enough for detailed examination have been fluted on one face. Most were discarded during the preparation for, or after the completion of, fluting of the second face. However, we assume the sequence of preparation for fluting was similar on both faces. Three aspects of fluting preparation are observable on the fluted preforms and channel flakes: facial preparation, tip preparation, and basal, or platform, preparation.

Facial preparation for fluting is visible to some extent on 11 items. Parallel-collateral flakes were removed at right angles to the longitudinal axis on the face to be fluted. These flakes were removed from a continuously ground and often slightly beveled platform prepared all along the lateral preform edges. The flakes were removed first along one edge and then the point was rotated so that a similar series of flakes could be removed from the same face along the other lateral edge (see Storck 1983: Fig. 2IIa). The removals from each edge terminated or met along the midline of the preform to form a well-defined medial "ridge" to guide the flute removals. Since there was a slight space between the platform segment used to remove each adjacent flake, the edges appear serrated. The tips of the serrations are simply the high points or remnants of the continuously prepared platforms between adjacent removals from an edge. This facial preparation generally removed any remaining evidence of scars from the preceding thinning stage although a remnant facet of this earlier stage is present on one face of the largest complete point (Fig. 20a) just beyond the flute scar.

Tip preparation for fluting varies. On one tip end (Fig. 22h) there does not appear to have been any preparation. This biface has the serrated edges right to a pointed tip. However, there is no evidence of subsequent attempts at fluting the prepared face and it seems possible that the tip would have been blunted prior to this fluting if the manufacturing sequence had been continued. Two other preforms (Fig. 22d, e) have pointed tips but undoubtedly they are a product of attempts to finish the point after successful fluting on both faces.

The remaining five preforms with intact tips have blunt ends. In one case, this blunt end is apparently a platform remnant of the original flake blank while in another, it is a flat surface. It seems possible that this flat surface is a remnant of the flat "bottom" of the core from which the flake blank was detached. Similar "bottom" remnants at preform tip ends have been reported from Folsom (Tunnell 1975:10–11). Two other tips are blunted by flaking (Fig. 24a, b). Both are also lightly ground. The final tip was also blunt but the exact nature of the blunting cannot be determined because of slight breakage due to collapse during the flute removal.

Another notable feature of the tip end preparation is found on four items. In these cases, the tip has been deliberately narrowed along both lateral edges for a short distance from the distal extremity. This results in a distinct break in outline (see, especially, Figs. 22e, 24a). Slight remnants of such narrowing also appear to be present on two preform tips which were altered subsequent to fluting (Fig. 22a, d) as well as on an unresharpened fluted point (Fig. 20a). The function of this preparation is debatable. However, we suggest this indicates the preforms were placed in some sort of holding device during fluting which extended slightly down the lateral edges of the tip. As such, it would be the first positive evidence of a holding device ever reported from fluted point industries.

Basal preparation for fluting is not well preserved in the fluted preform assemblage. However, the channel flakes (Chapter 5) clearly indicate the use of a well-prepared, ground and isolated platform. A slight bevel was prepared across the base of the face to be fluted. The platform was formed in the center of this beveled edge by additional flaking.

Several breaks or errors produced during fluting are represented in the assemblage. Two snapped tips are presumably the result of the collapse of the preform under the pressure of the detaching flute blow. Two specimens also indicate the removal of the tip end by

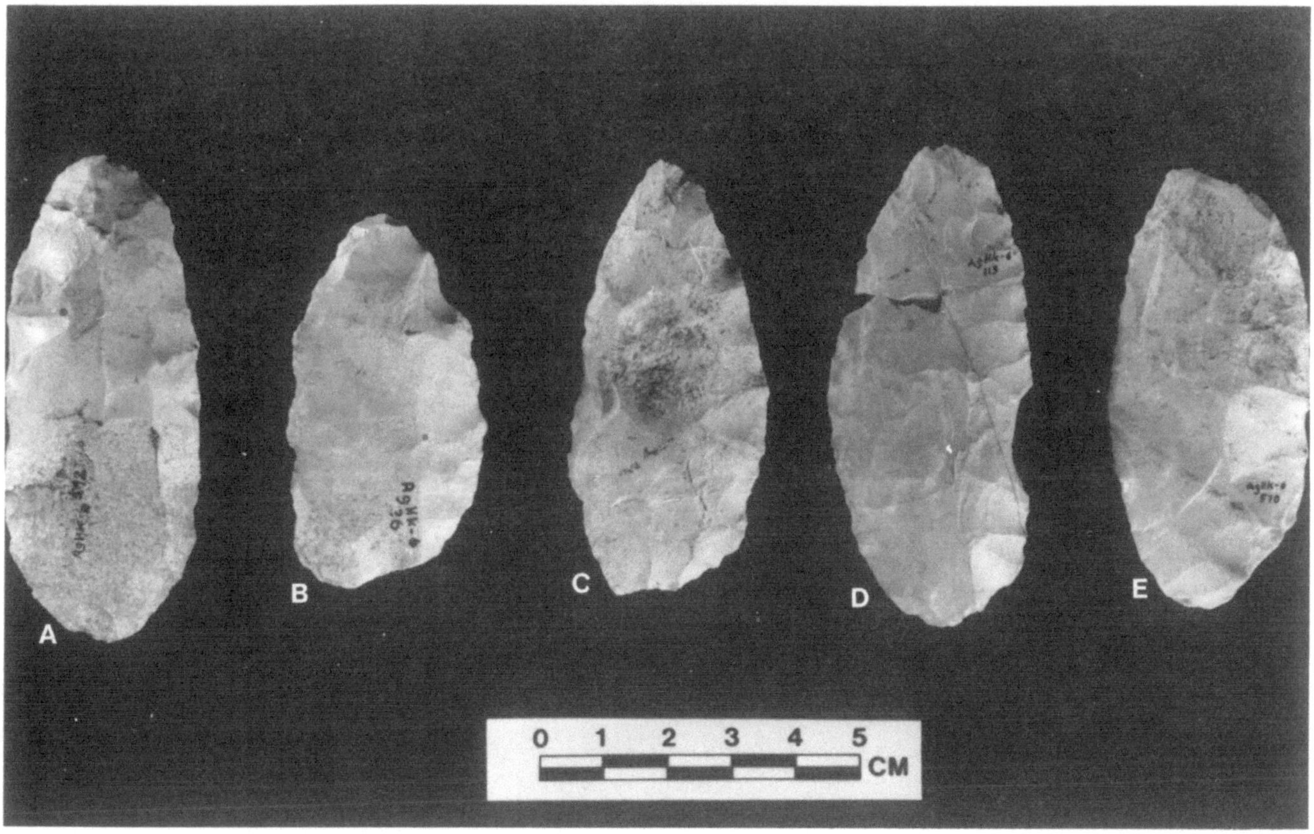

Figure 26. Unfluted preforms (**a-c**) and small oval bifaces (**d-e**).

an *outre-passe,* or plunging flute.[1] In one case (Fig. 24a) it has been possible to fit two channel flake segments together. The resulting 25 mm segment can then be attached to a tip fragment and it is clear that this channel flake removal detached the tip. This tip/channel flake is on a distinctive piece of Collingwood chert and an almost complete channel flake of the same distinctive coloration was recovered from the same area of the site (A-west). This probably represents the removal of a flute from the other face of the preform. The preform base was not recovered and it seems probable that it was finished into a small complete point and then, removed from the site. In other words, the small size of the removed tip end did not prevent finishing of the basal section into a point.

The reworking of shorter preforms produced by outre-passe terminations into smaller points is demonstrated conclusively by the second example. This preform was successfully fluted on one face and the distal segment of the channel flake removed was also recovered. The flute on the second face was markedly off-center and resulted in a plunging flute when the lateral edge was encountered. A channel flake segment removed in this operation can be reattached to the tip. In turn, the underside of the channel flake can be replaced onto a small preform with a pointed tip (Figs. 27, 28). Clearly, after the plunging flute, the knapper attempted to make the short basal section into a small point by pointing and finishing the tip end. However, during this process, a large segment (also recovered) was removed near the tip on one face due to a material flaw and the preform was discarded.

Two other probable errors related to fluting should be mentioned. As noted earlier, a ridge was created down the midline of the preforms by lateral flake detachment terminations in order to provide a ridge to guide flute removals. This ridge is very pronounced on one item (Fig. 22a). It would appear that the pronounced ridge resulted in the production of a very short, deep, broad flute that bit too deeply. The base of this preform exhibits an old break. It seems possible

[1] An error in flake removal resulting in the removal of a large section, instead of a more gradual feathering.

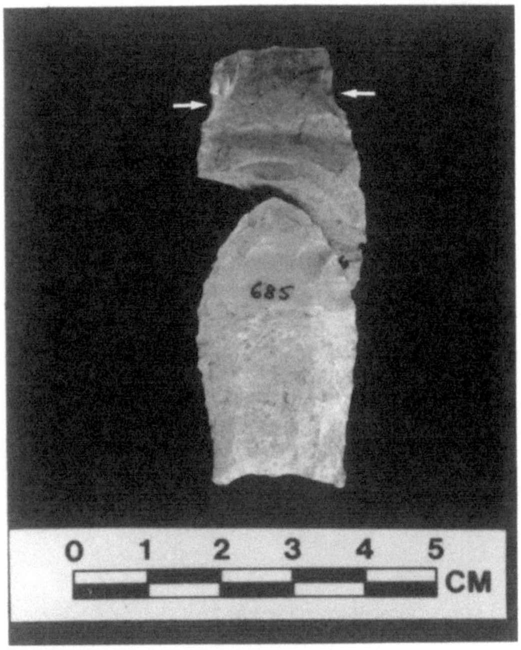

Figure 27. Outre-passe tip replaced on fluted preform, obverse view.

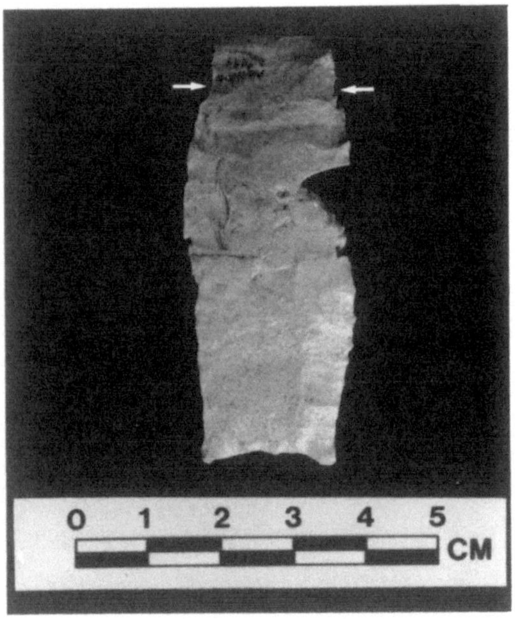

Figure 28. Outre-passe tip replaced on fluted preform, reverse view.

that the deep flute removal weakened the base causing it to break during attempts at finishing.

A second preform with breakage related to fluting is atypical because of its very small size and its fluting from the tip end on one face (Fig. 22i). This fluting from the tip end is unique to this item among known Barnes points. The tip flute terminated halfway down the face towards the base. In order to thin the base beyond the extent of the tip flute, a deep thinning flake was removed from one edge which successfully thinned a midsection area closer to the base. The knapper then attempted the same operation adjacent to the base to thin this area. This was unsuccessful and the preform split by a perverse fracture (a different interpretation is presented in Ellis 1984:138). We propose that this item is a product of a novice or unskilled knapper.

After successful fluting, the base was finished by the Barnes finishing flake removals. Two preforms discarded during the fluting of the second face, suggest the base could be finished in this manner on one face prior to even attempting flute removals on the second face. After the base was thinned, the tip was finished, the lateral edges were regularized by a minimal amount of selective marginal retouch and the lateral basal edges were ground.

External Comparisons

At least four fluted point types have been suggested in the eastern Great Lakes area: Enterline, Gainey, Barnes and Crowfield (Deller and Ellis 1984; Ellis 1984; Roosa 1963, 1965; Roosa and Deller 1982; Simons et al. 1984; Storck 1984a, 1984b; Wright 1981a, 1981b). These could be characterized respectively as: Clovis-like, Bull Brook-like, Cumberland-like and Reagen-like. The Enterline type is poorly known and controversial and so will not be considered here. The other three types are relatively well established because of their occurrence on sites yielding substantial point samples. Although the published sample from the Gainey site is small (Simons et al. 1984), sufficient data have been published on the points or can be inferred from illustrations to show their distinctiveness from Barnes and Crowfield points. Examples of the three point types are shown in Figures 29 to 31.

These types are generally thought to represent a temporal series rather than contemporaneous variation (see Deller 1988, 1989; Deller and Ellis 1988). There is considerable overlap in the spatial distribution of the types, yet, in the vast majority of cases, individual sites yield only points of one type. In short, there is no regionalization of the types to the extent which might be expected

Figure 29. Gainey points. **a**: Fernhill, Ontario, Upper Mercer chert; **b**: Uniondale, Ontario, Collingwood chert; **c**: Thedford area, Onondaga chert.

if they were measuring solely contemporaneous "social" variation. This can be seen in the distribution of known sites or findspots in southern Ontario (Figs. 32–34). As well, the three types or comparable forms occur over large areas of the adjacent United States (see Deller and Ellis 1988; Payne 1982; Roosa 1965; Simons et al. 1984). Also, even the largest sites, such as Parkhill (Roosa 1977a) and Fisher (Storck 1983), which are the best candidates for aggregation sites, yield only points of one type. Again, this suggests they are not monitoring contemporaneous social distinctions.

Neither are the types the product of the raw materials on which points are made. Examples of all of the types can be found made of the same chert type, such as Collingwood (Figs. 32–34). This is not meant to imply that raw material does not account for any variation. For example, as noted earlier, Barnes fluted point length on unresharpened forms seems to be correlated to some extent with raw material. While all types can occur on certain raw materials, overall the specific sources and frequencies thereof used for point manufacture vary significantly between types. These data indicate marked differences between the groups producing each type in terms of the sizes of the annual ranges covered, the specific areas and directions of movement over annual rounds, and the directions of interactions with groups in surrounding areas (see Deller 1989; Shott 1986). These data strongly suggest that each type was produced by groups with quite different cultural systems.

We should stress too, because it has engendered much discussion, that the different types are *not* a product of resharpening or reworking the same original form. This reworking has always been carefully evaluated in the definition of each point type (see Roosa

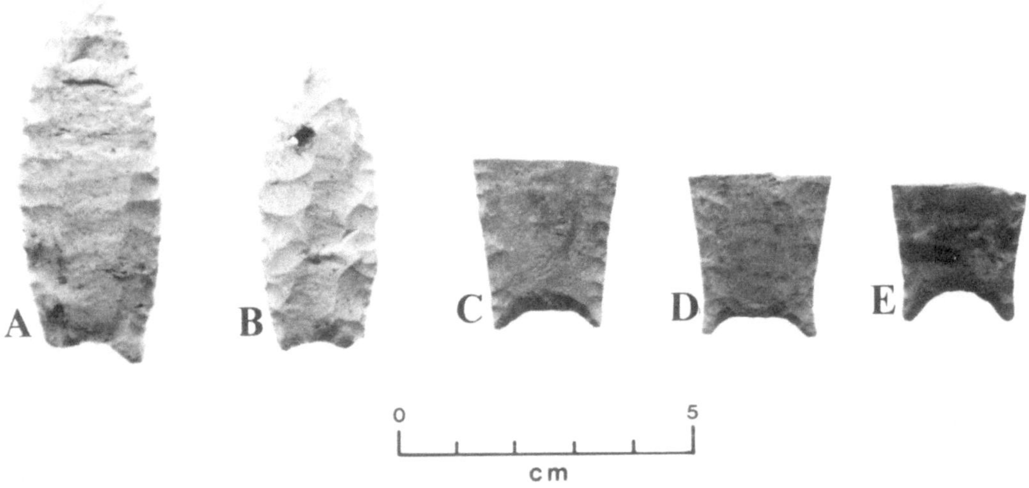

Figure 30. Barnes points/preforms. **a**: preform, Collingwood chert, Thedford II site; **b**: point, Collingwood chert, Fisher site; **c**: point, Bayport chert, Thedford II site; **d**: point, Bayport chert, Mullin site; **e**: point, Onondaga chert, Parkhill site.

1965). In short, attributes markedly affected by this factor, such as length, have not been used to create the types. However, variation in the way a point is resharpened can serve to distinguish some types. Crowfield points, for instance, are resharpened in a manner that produces straight, converging foresection tip edges and the overall resulting shape is pentagonal. The other types are never resharpened in this manner. This is not to deny that the reworking of some points, particularly the rare basal reworking, can obscure the type assignment of particular isolated examples (see, for example, the discussion of point width below). However, a consideration of all of the attributes and variable ranges used to isolate each type usually serves to distinguish these examples.

This brings us to an important consideration: these point types are "polythetic" ones (see Clarke 1968:37–38), defined on the basis of a large number of characteristics. In sum, as we perceive them, and following a trend initiated by Wright (1981a), no single characteristic, be it an aspect of fluting, a particular measurement of outline shape, etc., is sufficient to assign a particular point to a type. Rather, a point is assigned to a type on the basis of the consistent co-occurrence of a large percentage of the defining characteristics. For example, Figure 20e has short flutes in contrast to the long flutes characteristic of most Barnes points. Nevertheless, it is a Barnes point because it exhibits a large number of the other characteristics of the type to be outlined below such as a narrow basal width (under 20 mm), single fluting, fishtails and maximum width around midpoint. Furthermore, the *association* of particular attribute states and variable ranges is important in assigning particular points to a type because more than one type can exhibit a certain characteristic. For example, both Barnes and Crowfield points have narrow bases (under 20 mm). However, they differ in many other characteristics associated on the same point and narrow bases do serve to distinguish both of these types from Gainey points.

Finally, given that the types are monitoring temporal variation, it seems that each represents a "slice" of a continually evolving system. In other words, the types represent an arbitrary segment in a temporal continuum of morphological and technological change, sufficiently separate to isolate a different type. We suspect that this accounts for the majority of the known Ontario fluted points which cannot be easily assigned to certain types (i.e., they appear somewhat intermediate between types). For example, some points (Deller 1979: Fig. 5–29a; Garrad 1971: #16) appear to fall between Gainey and Barnes points and may be intermediate in time between the two. Similarly, some points appear intermediate between Barnes and Crowfield points (Deller 1976a: Plate 3v). It is worth noting that we have not seen points which are both Gainey-like and Crowfield-like. This is to be expected in a temporal series, as these are probably the earliest and latest in the sequence, respectively. In short, the intermediate forms are Barnes points.

Similarities between the Thedford II points and

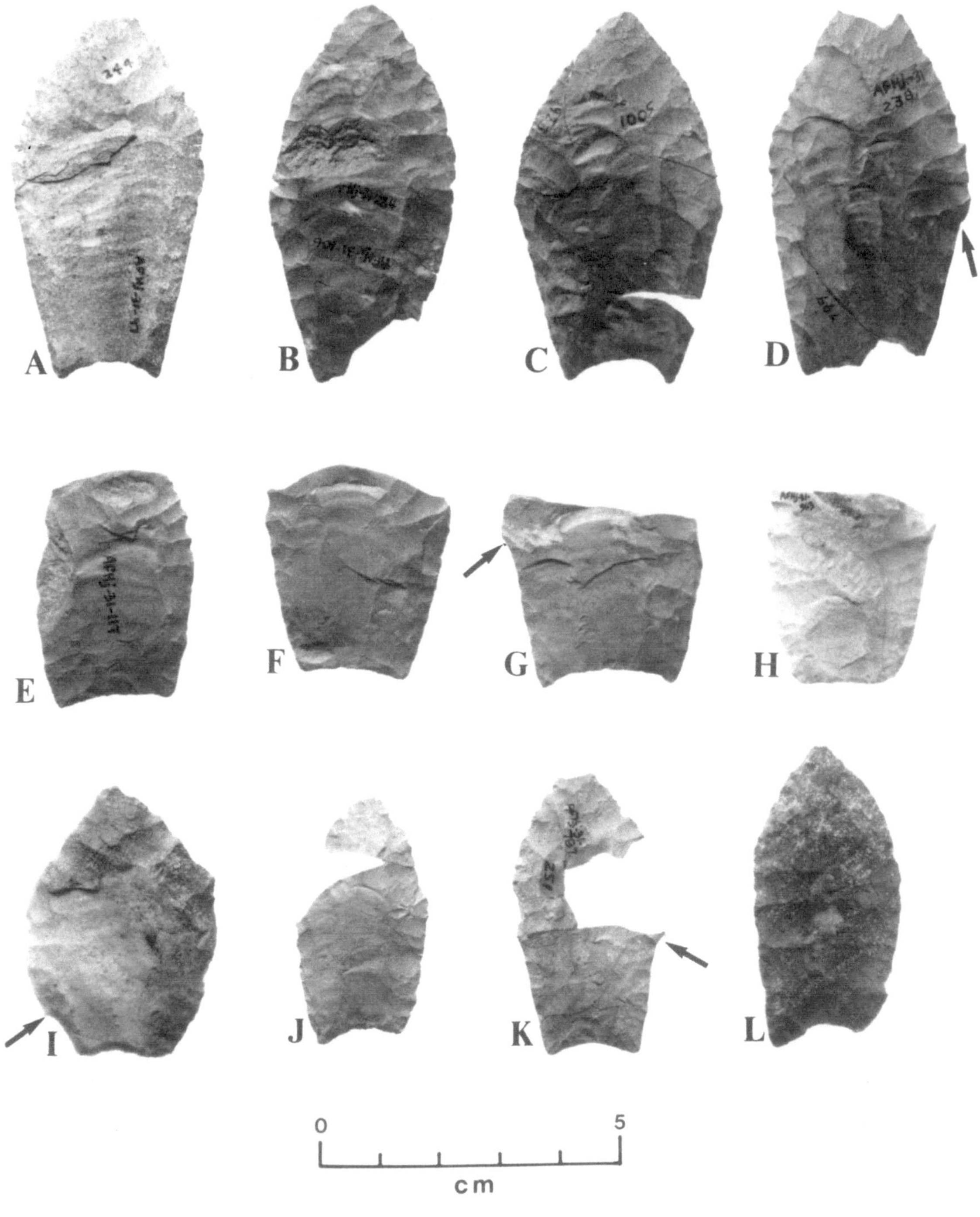

Figure 31. Crowfield points and Holcombe point. **a-k**: Crowfield points, Crowfield site. Arrows indicate shouldered examples. **l**: Holcombe point, Tedball site.

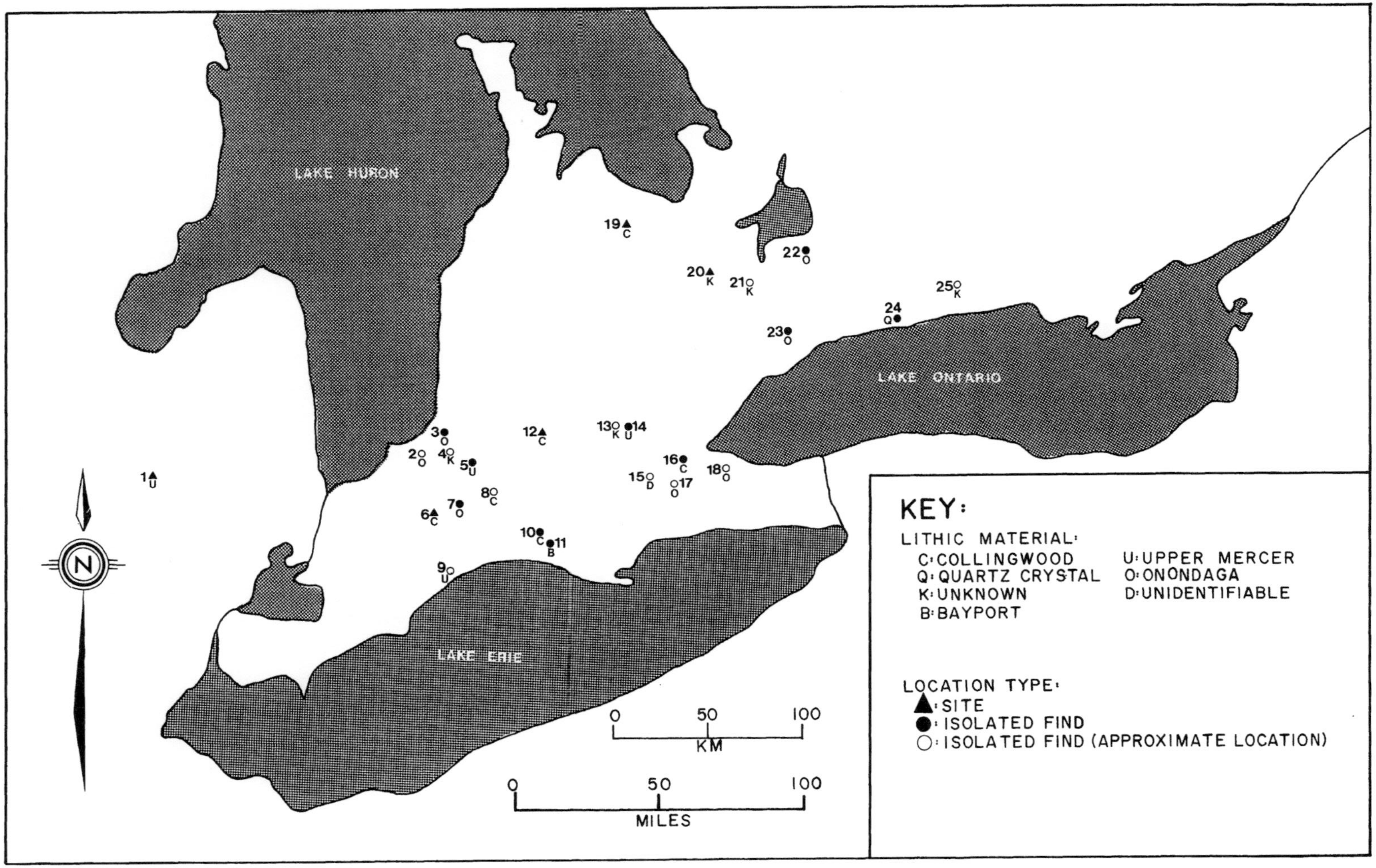

Figure. 32. Gainey complex sites and findspots. **1:** Gainey site (Simons et al. 1984); **2:** Thedford area (Fig. 29c, this report); **3:** Deller 1976b: #A8; **4:** Garrad 1971: #21; **5:** Garrad 1971: #20 (Fig. 29a, this report); **6:** Garrad 1971: #15, Deller 1979: #41; **7:** Garrad 1971: #18; **8:** Museum of Indian Archaeology, London, Reference #979–9–734783; **9:** Garrad 1971: #13 (corrected location); **10:** Garrad 1971: #27; **11:** previously unreported; **12:** Uniondale site (see Fig. 27b, this report; James Keron, pers. comm.); **13:** Garrad 1971: #39; **14:** previously unreported; **15:** Garrad 1971: #33, Kidd 1951; **16:** Garrad 1971: #36; **17:** Garrad 1971: #35; **18:** F. Wood collection, McMaster University, Reference #9347; Dr. W.C. Noble, pers. comm.; **19:** Kolapore site (Storck 1984b: Fig. 14); **20:** Banting site (Storck 1979); **21:** Garrad 1971: #42, Kidd 1951; **22:** Storck 1982: Fig. 7h; **23:** Garrad 1971: #43; **24:** Newcastle fluted point (Roberts 1984: Plate 5); **25:** Garrad 1971: #45.

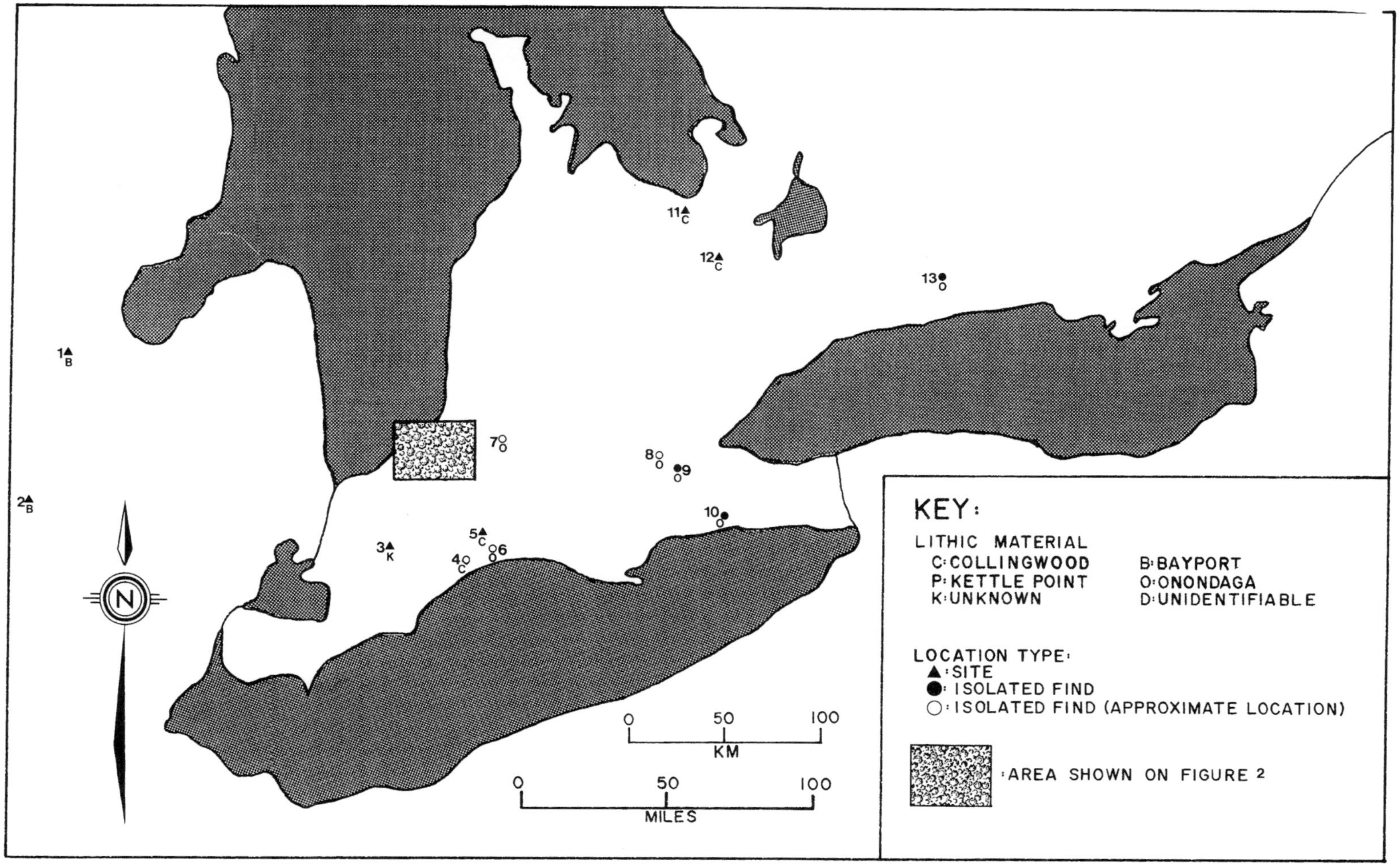

Figure 33. Parkhill complex sites and findspots. **1:** Barnes site (Wright and Roosa 1966; Voss 1977); **2:** Leavitt site (Shott 1986); **3:** Babula Farm (Ian T. Kenyon: pers. comm.); **4:** Garrad 1971: #14 (corrected location); **5:** Mullin site (Deller 1979: #51, see Fig. 30d this report); **6:** Garrad 1971: #25, Kidd 1951; **7:** Garrad 1971: #23; Kidd 1951; **8:** Garrad 1971: #34; **9:** Glass site (William Marshall: pers. comm.); **10:** Garrad 1971: #31; **11:** Fisher site (Storck 1982: Fig. 1, #3, 1983); **12:** Banting site (Storck 1979); **13:** Garrad 1971: #44.

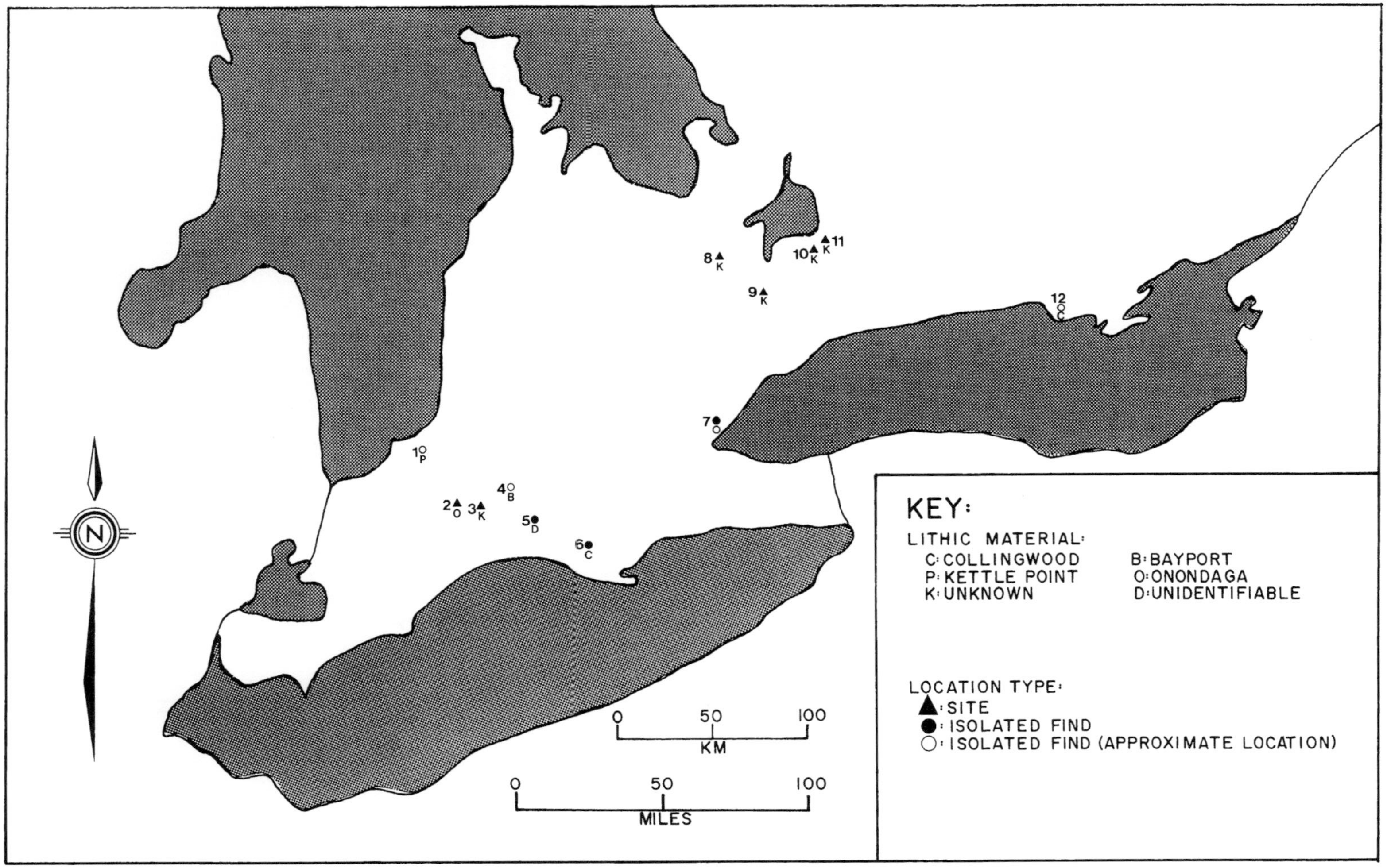

Figure 34. Crowfield complex sites and findspots. **1:** Bass point; previously unreported; **2:** Crowfield site (Deller and Ellis 1984); **3:** Deller 1979: #60; **4:** Garrad 1971: #24; **5:** T Creek Site (James Keron: pers. comm.); **6:** Rice, Deller 1988; **7:** Roberts 1984: Plate 1; **8:** Hussey site (Storck 1979: Plate 16c); **9:** Zander site (Storck 1982: Fig 7b; Stewart 1984); **10:** Udora site (Storck 1982: Fig. 7f); **11:** Watpool A site (Storck 1982; Fig. 7g, 1984a: Plate 3, right); **12:** Garrad 1971: #46 (see Roberts 1984: Plate 7).

Barnes points from other sites are outlined below. Some criteria are partially redundant (see for example, the discussion of point maximum width below). However, each criterion is worth emphasizing because it can be useful for distinguishing the type affiliation of fragmentary points. In making comparisons, we have been fortunate in having access to point samples from several major sites, including Fisher, Parkhill and Crowfield. For Gainey points, we have analyzed mainly isolated finds in various private collections and institutions. We have used only relatively complete points of this type in making comparisons. In the case of isolated finds of both Gainey and the other point types where the measurements have not been previously published and there appears to be little chance of such data being published in the future, we have included the measurements and other pertinent data for each individual point in Appendix C. Our sample consists solely of finished points and not preforms, as the inclusion of preforms would markedly skew the samples, particularly with regard to continuous variables. Preform data have been included only for measurements such as the banding orientation on Collingwood chert points where it is obvious no skewing will occur. The data on the various point samples are summarized on Tables 12 to 17.

In the following, we note key similarities between the Thedford II samples and Barnes points. Continuous variables are considered first.

Continuous Variables

Lateral Basal Edge Orientation. Wright (1981a) has demonstrated that this orientation can be very useful for distinguishing the point types. The degree of expansion or contraction of the edges from the base is measured on a 180° scale. This measure is referred to as *face-angle.* It consists of the angle between a line drawn across the base and one along the lateral basal edge ignoring ear-flaring or fishtails (Figure 25). The measure is taken for both basal corners and averaged for each point if both corners are complete. Otherwise, the measure for the single complete edge is used. Measurements of 90° indicate the sides are roughly parallel while those below 90° contract and those above 90° expand.

The measurements for Barnes points and the present sample cluster tightly around 95–100° indicating moderately expanding lateral edges from the base. When compared to the other types, a clear trend emerges from the more parallel-sided Gainey points to the slightly expanding Barnes points to the markedly expanding Crowfield points. This highlights the intermediate posi-

TABLE 12
Length, Fluted Point Samples

Sample*	N	R	X	S	C.V.
G2	8	33.6–94.5	59.8	18.887	31.58
B1	4	49.5–77.5	57.4	12.145	21.15
B2	3	88.0–105	94.5	—	—
B3	5	37.1–52.2	45.6	6.373	13.98
B4	0	—	—	—	—
B5	4	34.8–49.2	43.1	6.423	14.92
C	7	41.1–63.8	54.7	8.947	16.35

*Samples: G2 = Gainey isolates; B1 = Barnes isolates; B2 = Barnes, Thedford II; B3 = Barnes, Parkhill site; B4 = Barnes, Barnes site; B5 = Barnes, Fisher site; C = Crowfield site. See also Table 17.

TABLE 13
Width, Fluted Point Samples

Samples*	N	R	X	S	C.V.
G2	14	23.2–30.0	26.1	1.799	6.90
B1	8	19.0–26.5	22.3	2.694	12.06
B2	5	22.0–27.2	25.5	2.158	8.46
B3	21	11.1–24.6	20.4	3.020	14.79
B4	0	—	—	—	—
B5	18	10.4–22.4	17.0	2.945	17.30
C	16	22.3–35.4	30.8	3.590	11.66

*For sample key see Tables 12 and 17.

TABLE 14
Thickness, Fluted Point Samples

Samples*	N	R	X	S	C.V.
G2	13	6.0–8.5	7.1	0.807	11.29
B1	10	4.7–7.3	6.0	1.035	17.19
B2	6	5.2–7.5	6.1	0.945	15.49
B3	26	4.6–7.9	5.6	0.835	14.91
B4	3	5.0–6.0	5.3	—	—
B5	16	4.2–6.6	5.3	0.648	12.14
C	19	3.2–5.7	4.6	0.667	14.44

*For sample key see Tables 12 and 17.

TABLE 15
Basal Width, Fluted Point Samples

Samples*	N	R	X	S	C.V.
G2	14	22.0–28.0	25.2	1.786	7.09
B1	4	16.0–21.0	18.9	2.099	11.10
B2	4	16.5–19.5	18.4	1.350	7.34
B3	25	15.0–19.2	17.0	1.194	7.04
B4	5	17.0–20.0	18.4	1.140	6.20
B5	20	13.8–18.6	15.1	1.398	9.28
C	13	13.1–22.8	17.9	2.540	14.18

*For sample key see Tables 12 and 17.

TABLE 16
Basal Concavity Depth, Fluted Point Samples

Samples*	N	R	X	S	C.V.
G2	13	3.5–8.0	5.5	1.274	23.25
B1	7	2.0–5.2	3.5	1.161	33.08
B2	5	3.0–5.0	3.6	0.820	23.03
B3	40	2.1–5.7	3.8	0.971	25.55
B4	0	—	—	—	—
B5	20	1.6–4.7	3.2	0.879	27.71
C	13	0.4–4.2	2.1	1.185	56.97

*For sample key see Tables 12 and 17.

TABLE 17
Key to Figures and Tables, Fluted Point Samples

SAMPLES

G1 = Gainey site (based on Simons et al. 1984: Fig. 8b).

G2 = Gainey point isolated finds (based on measurements of points in institutional and private collections; sample includes Garrad 1971: #'s 13, 15, 20, 27, 33, 35; Figs. 21b and c, this report; Wright 1981b).

B1 = Barnes points isolated finds (based on points in institutional and private collections; sample includes: Garrad 1971: #'s 14, 34; Deller 1979: Figs. 4–10, 4–20, 5–51 and others).

B2 = Barnes points, Thedford II site.

B3 = Barnes points, Parkhill site (based on sample of points from that site measured by Ellis).

B4 = Barnes points, Barnes site (based on Wright and Roosa 1966: Table 2, artifact 15, 16, 19 and Voss 1977: Table 1; Only measurements which appear to characterize complete finished points were employed).

B5 = Barnes points, Fisher site (based on sample of points in collections at Royal Ontario Museum measured by Ellis, courtesy of Dr. Peter Storck).

C = Crowfield points, Crowfield site (based on materials from the first season of excavation reported in Ellis 1984: Appendix C).

RAW MATERIALS

Unknown: If an isolated find, this designation means the point has not been personally examined by us to determine raw material. If a site, the dominant material is not known because of the small size of the collections or difficulties in sorting out the Paleo-Indian material from later occupations at the site.

Unidentifiable: This designation means the point in question has been examined by us but the raw material upon which it was made could not be determined.

tion of Barnes points and is highly suggestive of a temporal trend (Fig. 35).

Maximum Thickness. Data on this variable are summarized in Figure 36. Figure 36a shows the exact data plots while 36b plots data summaries in the form of mean, standard deviation and the 95% confidence limits of the mean. Basically, if the confidence limits (solid black bars) do not overlap, the differences are significant at the .05 level. Only the mean is shown for the Barnes site (B4; see Table 17) because of the very small sample size.

The Thedford II collection (B2) most closely approximates that of other Barnes sites. All Barnes samples with the exception of the Fisher site (B5) are significantly thicker than the Crowfield sample (C). Most Barnes point collections, and particularly those for which large sample sizes are available, are thinner than the Gainey sample representing isolated finds and small site collections in Ontario (G2). There is some degree of overlap which may be due to sample size. Moreover, if the point types represent arbitrary segments in a continually evolving form, perhaps some degree of non-significant overlap is to be expected. In other words, we believe that the overall trend towards thinner points from Gainey to Barnes to Crowfield is of significance even if there is some overlap between individual samples of each. Certainly, the end points (Gainey and Crowfield) are markedly different from one another.

It appears that thickness of Barnes points is affected by raw material. In particular, the thickest points in both the Barnes isolated finds and Thedford II sample (e.g., Fig. 20a) are on the "soft," mottled variety of Collingwood chert.

Basal Concavity Depth. Figure 37 plots basal concavity depths for the available samples. Only the mean for the Gainey sample is given as these depths were estimated from an illustration (Simons et al. 1984: Fig. 8b). Once again, the Thedford II sample approximates most closely the samples of other Barnes sites and the clear trend is for shallower concavities throughout the sequence of Gainey to Crowfield even though differences between individual samples are not always statistically significant. The sample from the Gainey site (G1) shows much more overlap with Barnes than the Gainey isolated finds. It is possible that other factors are influencing the Gainey site sample. In particular, two of the three points with shallow basal concavities at Gainey are on Ten Mile Creek chert while those with deep concavities are exclusively on Upper Mercer chert. As well, it is possible that the measurements from Gainey re-

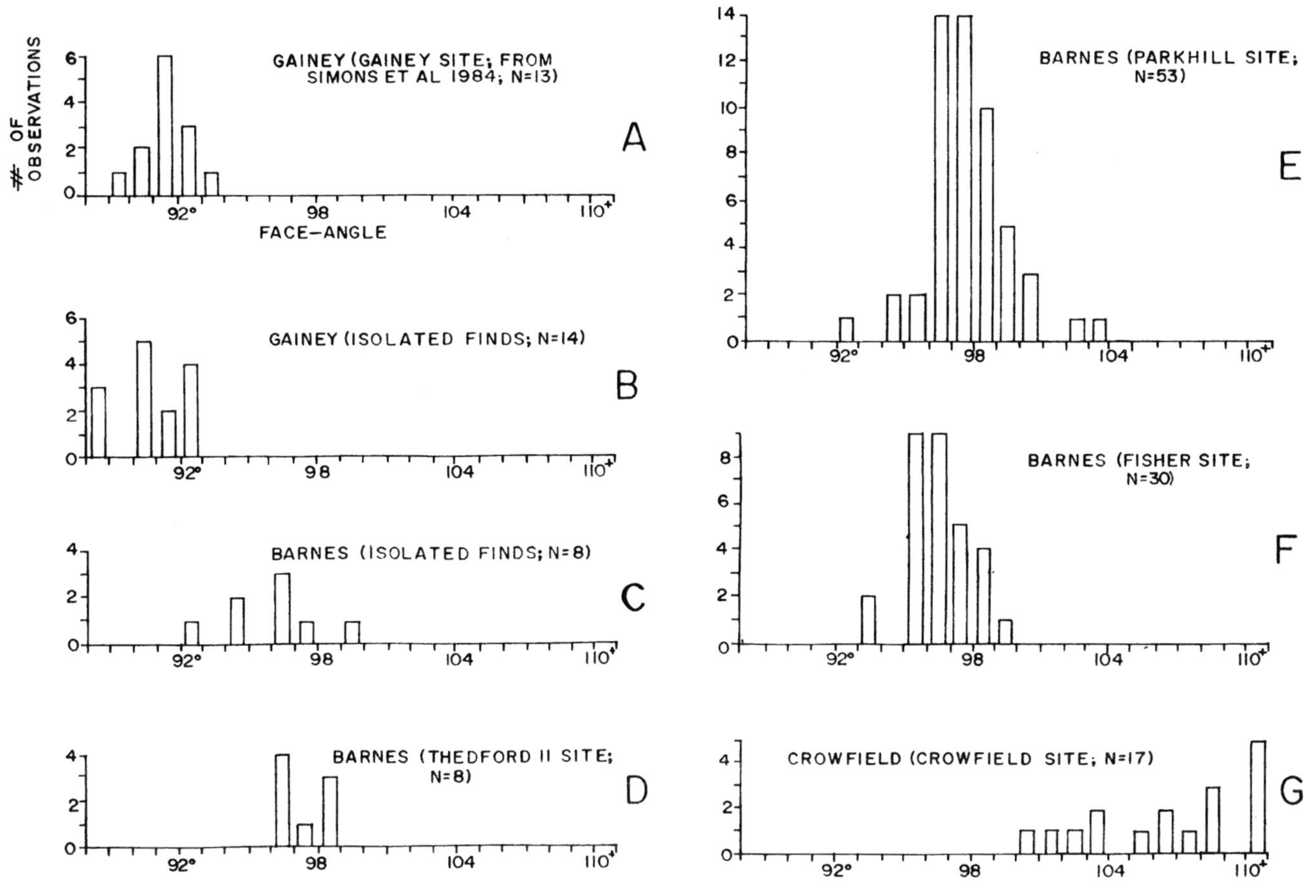

Figure 35. Face-angle measurements for fluted point samples.

ported by Simons et al. (1984) include the preform sample from this site. Finally, the pronounced ears on Gainey points are probably susceptible to more breakage and reworking than is the case for the other types, and thus, the production of shallower concavities. In any case, only Gainey points can exhibit deep basal concavities over 5 mm with any degree of regularity.

Maximum Width. Figure 38 compares this variable for the samples. Again, only the mean is given for the Barnes site collection due to the small sample size. Even though samples of each of the point types are quite distinct, Barnes points do not appear intermediate between the Gainey and Crowfield samples. Specifically, the Crowfield points are very wide even though, in line with previous trends, we should expect them to be the narrowest. We suspect this deviation is due primarily to one factor. Face-angle measurements indicate the Crowfield points markedly expand from the base. Generally, if one wishes to maximize this attribute, it is impossible to produce a narrower point. The more expanding a point is from the base, the wider it must be. In essence, because some characteristics are interrelated, a perfect ordering is impossible.

Another notable aspect of the width comparisons is the fact that the Thedford II sample (B2) overlaps considerably with the samples of Gainey points and appears significantly wider than the Parkhill site (B3) and Fisher site (B5) samples. This appears to be largely due to resharpening. As noted earlier, many of the complete, measurable points from Thedford II are "mint" and unresharpened which is a major contrast with the other Barnes samples. Since Barnes points expand from the base, we would expect length reduction through resharpening to result in a narrowing of the points as well. This also could account partially for the extreme

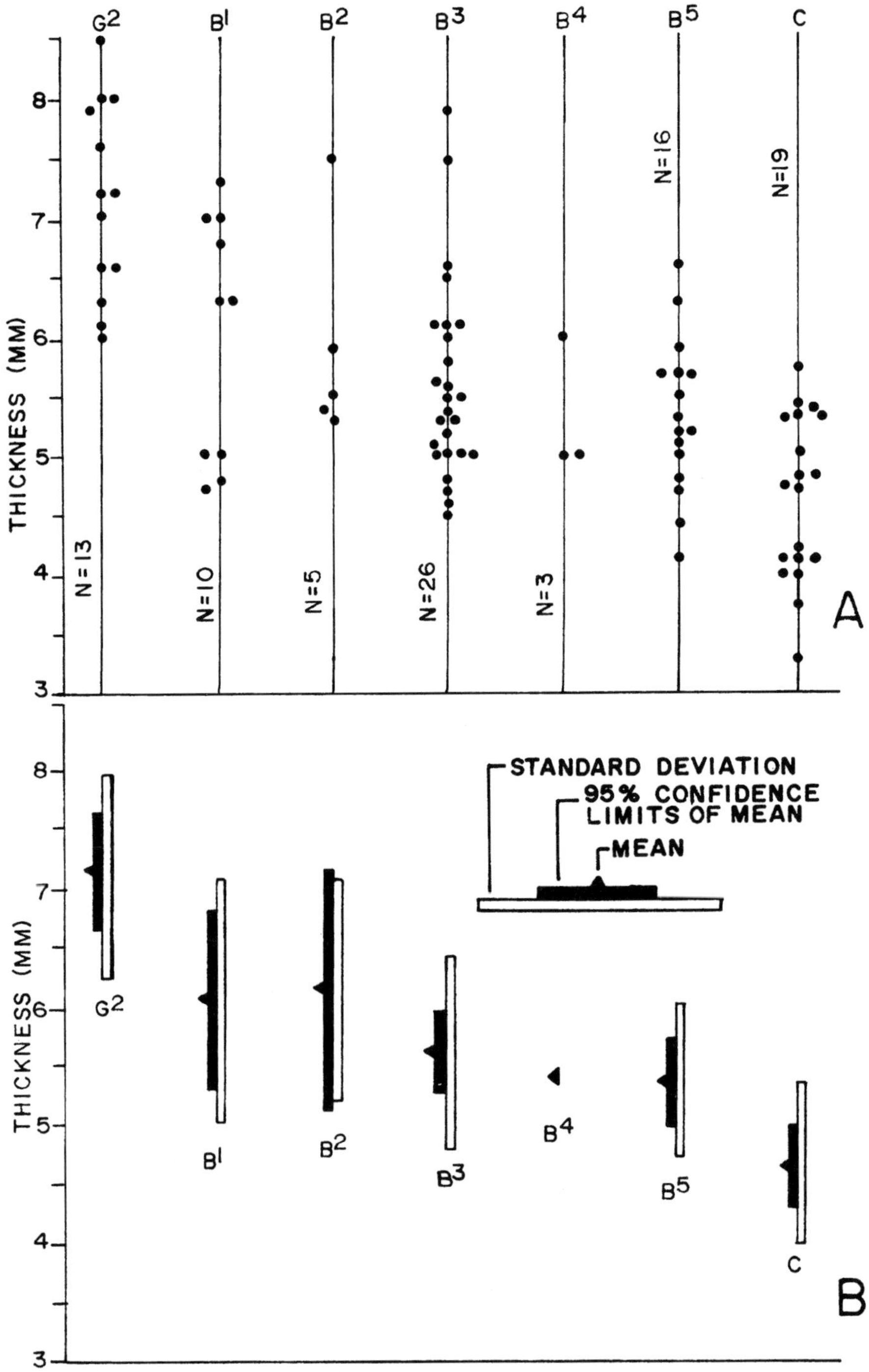

Figure 36. Maximum thickness, fluted points. [Reproduced from Deller and Ellis 1988 by permission of the publisher.]

width of the Crowfield points since our sample is from a probable grave cache and is probably less resharpened than points discarded in a normal manner.

Assuming length is an indicator of degree of resharpening, we should expect a decrease in Barnes point width as length is reduced. This trend will not be the case for Gainey points as they are essentially parallel-sided. Both inferences are confirmed if we plot length by width (Fig. 39). The Thedford II points are generally longer and thus, wider than the other Barnes samples. Hence, the difference in width between these points and some of the other Barnes samples are probably due to resharpening. However, the differences in width between the Fisher site (B5) and the other resharpened Barnes samples are not as easy to explain. These are so narrow that they are statistically different than even the Parkhill site sample. Since the Fisher site sample often tends to be the most different from the Gainey samples and/or the most similar to the Crowfield sample in terms not only of width but also, basal concavity depth, basal width (see below) and other characteristics, we would suggest this indicates a time difference with Fisher being later in time than the other Barnes sites.

Basal Width. As implied in previous discussions, this is a very useful criterion for distinguishing Gainey points from the other point forms. Gainey points almost exclusively have a basal width over 20 mm whereas the other types are under this size (Fig. 40). Unfortunately, basal widths are not yet available for the Gainey site. However, examination of drawings (Simons et al. 1984: Figs. 6 & 7) indicate the points at that site have a basal width predominantly and markedly over 20 mm wide. There is some variation among the Barnes samples. The Fisher site points are very narrow when compared to the Thedford II and Parkhill samples. In line with previous comments, this could suggest the Fisher site is later in time on the assumption the earliest Barnes points would have wider bases more closely approximating Gainey points.

Banding Orientation. As noted previously, there is a significant tendency for the Collingwood chert points in the present sample to have banding at right angles (0–5°) to the longitudinal point axis. This tendency is also evident at Parkhill and, to a certain extent, at Fisher (Fig. 41). The Fisher sample varies more than the other samples and this may again indicate a later time placement as the implied later Crowfield examples vary considerably in terms of this characteristic (Fig. 41e). The Gainey sample on Collingwood chert is too small for firm conclusions, but it appears to vary in the same manner as the Crowfield examples. These differences are strongly suggestive of different blank production and selection procedures between the Parkhill industry and the other industries.

Discrete Attributes

Position of Maximum Width. To a certain extent, this attribute is redundant with orientation of the lateral basal edges. For example, if lateral edges expand from the base, maximum width must occur above base. Nonetheless, it is notable that on Barnes points and the present sample, maximum width occurs just around midpoint or slightly below it. This contrasts with Crowfield points where maximum width occurs at or above midpoint (Deller and Ellis 1984:44; Ellis 1984:301).

Fishtails. These are definitely lacking on Crowfield points (Deller and Ellis 1984:44). The points from the Gainey site also lack fishtails (Simons et al. 1984). It is possible that this is due to the small sample size at Gainey. For example, the original sample from the Barnes site reported by Wright and Roosa (1966) lacked fishtails but subsequent excavations produced examples with quite pronounced features of this nature (Voss 1977: Fig. 4a). Some of the items from Ontario which we have characterized as Gainey points have fishtails but otherwise are identical or indistinguishable with present data from Gainey points (Garrad 1971: #15). Alternatively, it is also possible that the fishtailed Ontario points are closer in time to Barnes points than the Gainey site itself.

Be that as it may, all of the fishtailed points which we have called Gainey exhibit only slight ear-flaring as compared to the very pronounced corners seen on many Barnes points (Fig. 20g). Moreover, a notable aspect of the Gainey points is that the slight constriction or "waist" is much above the basal extremity so that the ears gradually flare, whereas on Barnes points, this constriction is immediately above the base and results in an abrupt flaring.

Pentagonal Outlines. As implied earlier, Crowfield points can be resharpened in a manner that produces a pentagonal outline. This is not found on the other point types or in the Thedford II sample.

Shoulders. The present sample lacks the single shoulder on one edge which is found on some Crowfield points (Deller and Ellis 1984:44–45). This indicates closer similarities to the Gainey and Barnes points.

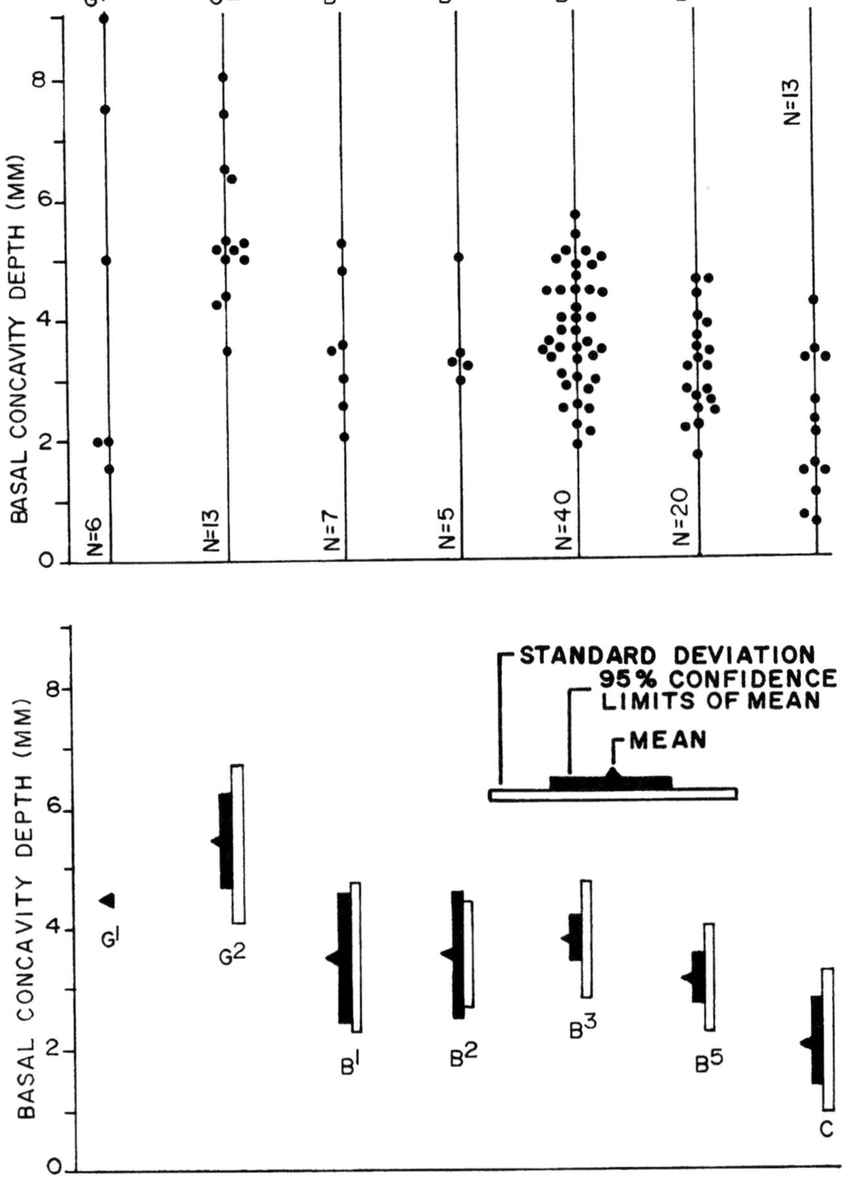

Figure 37. Basal concavity depth, fluted points.

Preform Tip Preparation. The present sample includes preforms with deliberate tip narrowing, perhaps to permit insertion into a holding device for fluting. This trait is known only from sites with Barnes points such as Parkhill (Deller 1980a: Plate C1). It is not known to occur on Gainey or Crowfield preforms.

Facial Preparation for Fluting. As noted earlier, a ridge was produced along the center of preform faces by collateral flaking. This ridge served to guide flute removals. This trait occurs on preforms from other sites with Barnes points (Storck 1983:83). It also occurs on Gainey points but does not occur on Crowfield points.

Number of Flutes. The present sample is identical to Barnes samples elsewhere in that most points exhibit single flutes and there are never more than two flutes to a face (Roosa 1977b: 91–92; Storck 1983:84). This contrasts

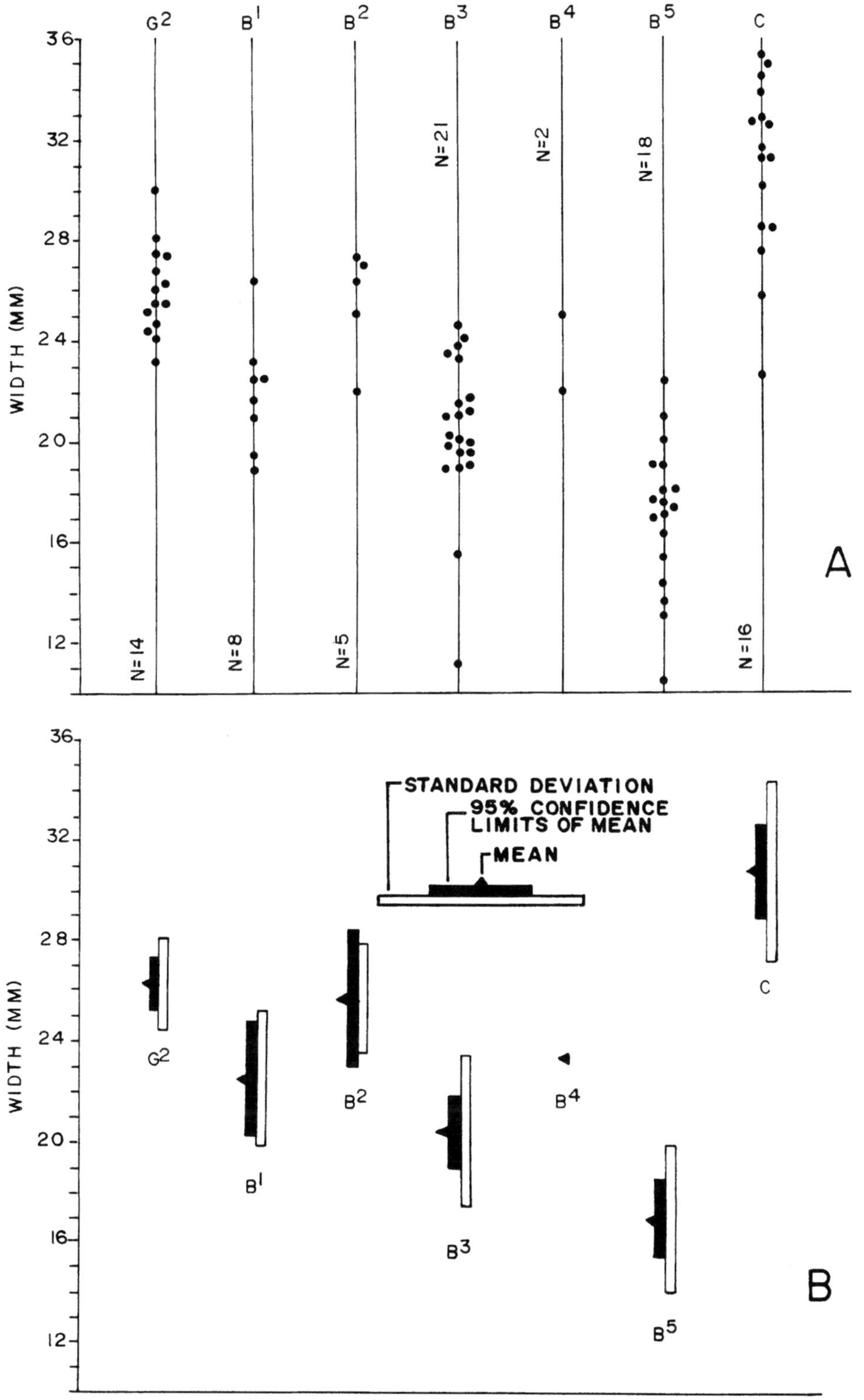

Figure 38. Maximum width, fluted points.

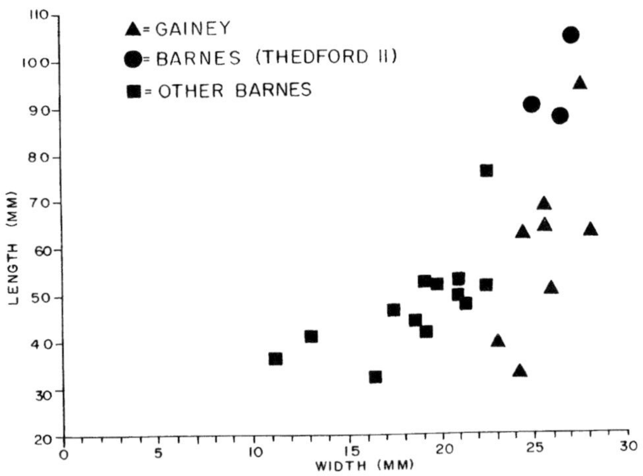

Figure 39. Plot of fluted point length by width.

with Crowfield points, which are often multiple fluted and can exhibit more than two flutes to a face (Ellis 1984: Tables 57, 59). Gainey points also tend to be single fluted but the occasional point can have three flutes on a face (Simons et al. 1984: Fig. 6d).

Flute Length. The present sample and Barnes points tend to have long flutes extending to or above three-quarters of point length on at least one face. In contrast, Crowfield points often have shorter flutes (see Ellis 1984:300) and we suspect this is the case with most Gainey points as well.

Basal Finishing Techniques. Both the present sample and other Barnes samples have the Barnes basal finishing technique. Generally they lack the fine, continuous retouch found across the basal concavity commonly found on types such as Folsom. Crowfield points lack the Barnes basal finishing technique and have the fine continuous concavity retouch. Gainey points commonly have Barnes finishing. Also, they often have the parallel retouch around the basal concavity on at least one face (Simons et al. 1984: Fig. 5e, f, g; 6c, d).

Summary

There are other contrasts between the various types but the above description is sufficient for our purposes. In sum, there is no doubt that the Thedford II points are Barnes points.

Other Bifaces

Concave-Base Bifaces

Two bifaces are small, lack lateral grinding and have concave bases (Table 18). One (Fig. 42a) is somewhat thick with a marked biconvex cross-section, sinuous edges, almost a "stemmed" appearance in plan and basal concavity grinding.

TABLE 18
Concave-based Bifaces

FC#	Length	Width	Thickness
37	44.4	19.0	5.7
320	40.7	23.6	8.0

FC# = field catalog number.

The second biface (Fig. 42b) is made on a thin flake. It is somewhat curved in longitudinal section and has a plano-convex transverse section. The plano face exhibits a facet of the original flake blank upon which the tool was made. There is short (under 15.5 mm) basal thinning on the convex face. This tool is similar to items from Debert (MacDonald 1968: Fig. 21a, b) referred to as "small hafted tools."

Alternately Beveled Bifaces

This type includes fragments of five large, well-made bifaces with high width to thickness ratios (Figs. 23e, 43; Table 19). Given their inferred size, it is possible that some were made on the nuclei of large biface cores. The sample includes two tip ends, two midsections and a basal segment.

TABLE 19
Alternately Beveled Bifaces

FC#	Material	Width	Thickness	Bevel Angle	
				1	2
33	Collingwood	38.7	9.5	40–50	45–50
29	Bayport	55.2	6.9	50–55	60–70
30	Bayport	—	8.7	50–65	55–65
473	Bayport	—	7.0	40–50	?
257	Bayport	—	6.1	35–45	?

FC# = field catalog number.

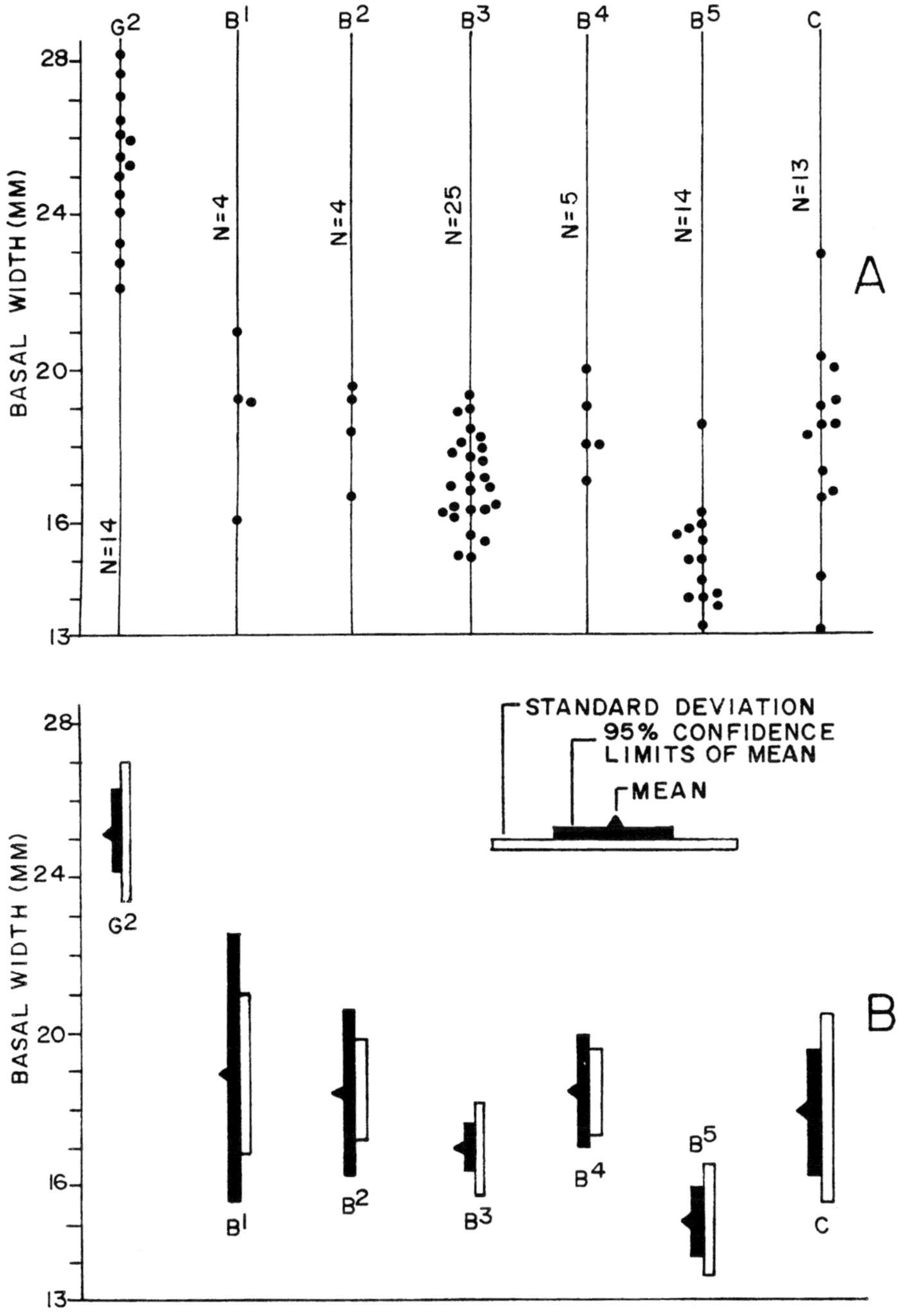

Figure 40. Basal width, fluted point samples.

Surfaces are covered by large, broad, flat, well-executed, thinning flake scars. Near the base, a fine intermittent retouch has been applied to regularize margins. The tip ends exhibit an extensive, unifacially retouched bevel on one face along one edge and along the other face on the other edge (i.e., an alternate bevel). This is probably a product of the continued unifacial resharpening of working edges. On those items which can be oriented with the tip away from the viewer in plan, the bevel occurs on the left edge of the visible face.

Excluding one tip end formed by a recent break, the fragments exhibit either one (2 examples) or two (2 examples) snap or perverse fractures at ends. On two items, retouch has been applied to these breaks which extends on one face back from the right-angled edge. This is highly suggestive of the recycling and reuse of these items as bend-break or snapped tools. One of these tools also exhibits a rounded juncture of a lateral edge and a snap formed by recycling. A final tool (Fig. 43d) has been recycled by retouching an end scraper bit and a notch into one end.

Large alternately beveled bifaces are a distinctive tool form which occurs in low numbers on several Parkhill industry sites such as Parkhill and McLeod (Roosa and Deller 1982:5). They are also known from the Hussey site (Storck 1979:53) and the Crowfield site (Deller and Ellis 1984:46) and are occasionally reported from western Folsom contexts (Frison and Stanford 1982:55; Irwin and Wormington 1971:30).

Small Oval Bifaces

Two complete bifaces are assigned to this type (Fig. 26d, e; Table 20). They are somewhat oval in outline and "twisted" in transverse section. Both have relatively broad thinning flake scars on each face. One lateral edge also exhibits a fine, continuous to intermittent bifacial retouch which serves to form a thin, slightly sinuous working edge. The opposite edge has a larger, gross retouch which forms a quite sinuous edge. In one case (Fig. 26d), a "bite" formed by edge collapse during flaking is found on this edge.

It is possible that these bifaces were originally intended to be point preforms but due to errors in reduction, such as an irregular cross-section, they were recycled into other uses. However, both exhibit diagonal banding which runs almost end to end. Given previous discussions, this lowers the possibility that they were preforms for projectile points.

These two bifaces are similar to what MacDonald (1968:79) refers to as "small bifacial knives" at Debert in that they have a finely retouched edge opposite a more sinuous, less finely flaked edge. However, unlike the Debert sample, there is no evidence of the original flake surfaces upon which the tools were made.

TABLE 20
Small Oval Bifaces

FC#	Length	Width	Thickness
84/113	75.4	32.3	8.6
570	68.7	31.0	12.1

FC# = field catalog number.

TABLE 21
Other Bifaces

Type	Length	Width	Thickness
Backed biface	71.0	30.9	7.2
Channel flake point	20.5	10.5	2.1

Other Bifaces

There are three other relatively complete bifaces as well as five small biface fragments (Table 21). Four of the latter are on Bayport chert and thus may be associated with later occupations on the site.

One of the more complete items may be a small *pièce esquillée* although it is certainly not typical of these items as reported from other sites. It seems more likely that it is simply a fragment of a narrow, thick biface.

A second biface is an example of a distinctive tool form, which we refer to as *backed bifaces*. These tools occur at several Paleo-Indian sites in Ontario such as Parkhill, Dixon, Wight and Crowfield (Ellis and Deller 1988; Deller and Ellis 1984:46). The tool (Fig. 44) has a flat back formed by an unflaked weathered surface along one lateral edge. This back has been used as a striking platform to remove broad, thinning flake scars over both surfaces. The weathered back, and the right-angled banding, indicate this tool was made on a top-corner struck blank.

The tip end and the edge opposite the back has a fine, bifacial flaking. The lateral edge exhibits a slight bevel and is highly exfoliated and rounded suggesting intensive use. The base exhibits a broad (18 mm), shallow (2.5 mm), steeply retouched concavity. The presence of one or two such concavities is a consistent feature of Parkhill industry backed bifaces. It serves to distinguish them from those backed bifaces known to be associated with the Crowfield industry (Ellis 1984).

The final biface (Fig. 42c) is a small, complete point made on the basal section of a channel flake. These artifacts are known at other Parkhill industry sites such as the type site (Roosa and Deller 1982:5). They have been suggested to be a defining characteristic of this industry since they have not yet been reported on sites

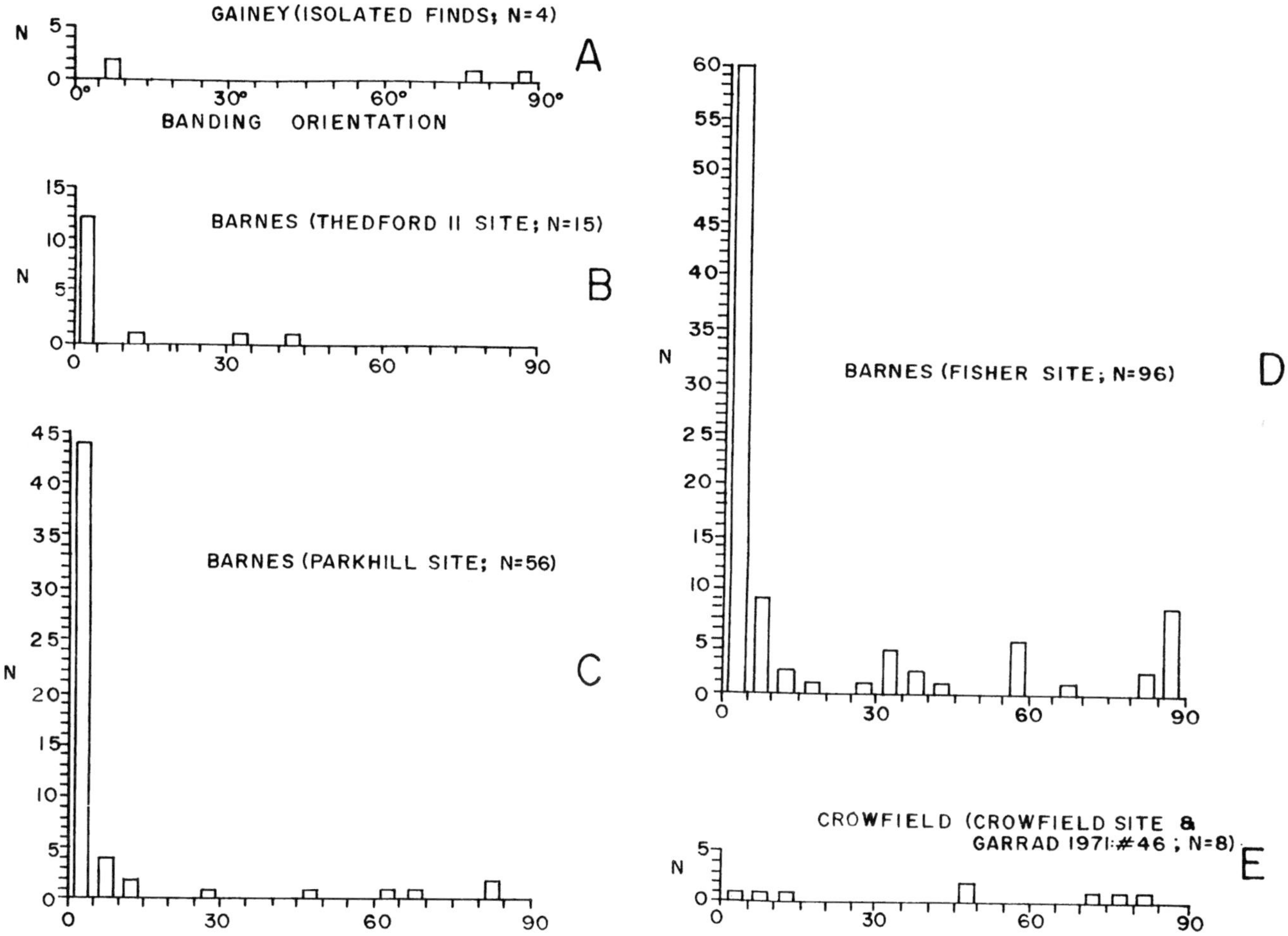

Figure 41. Banding orientation, Collingwood chert fluted bifaces.

associated with other fluted point industries in the Great Lakes area. The Thedford II point tip's apex is a remnant of the naturally pointed isolated striking platform used for flute removal. A very short (1–2 mm) retouch has been applied on both faces along the edges to refine outline shape and produce a basal concavity.

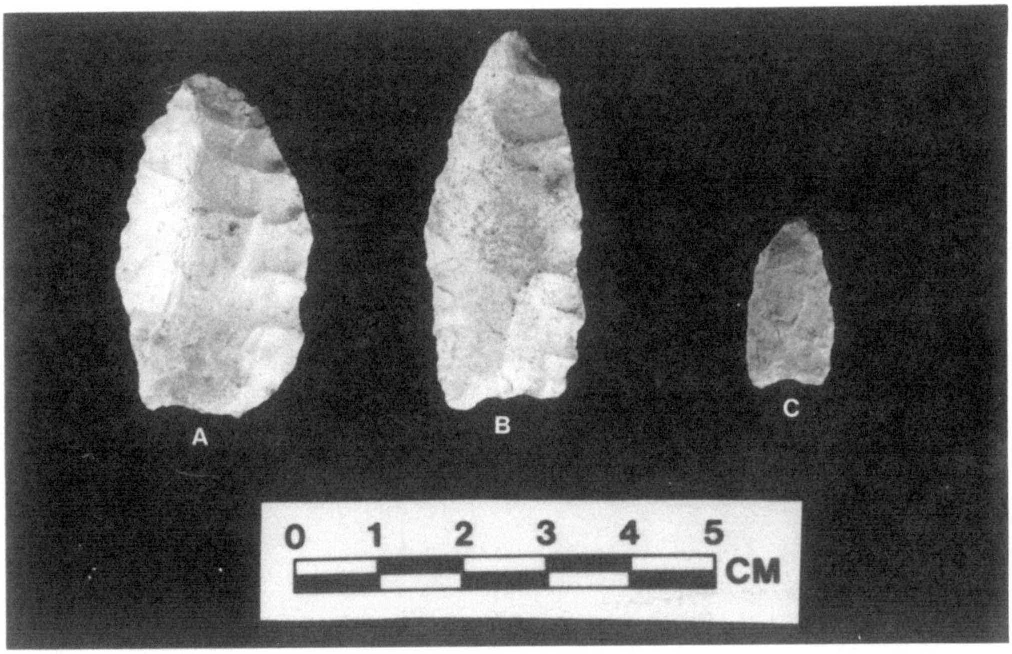

Figure 42. Concave-base bifaces (**a-b**) and channel flake point (**c**).

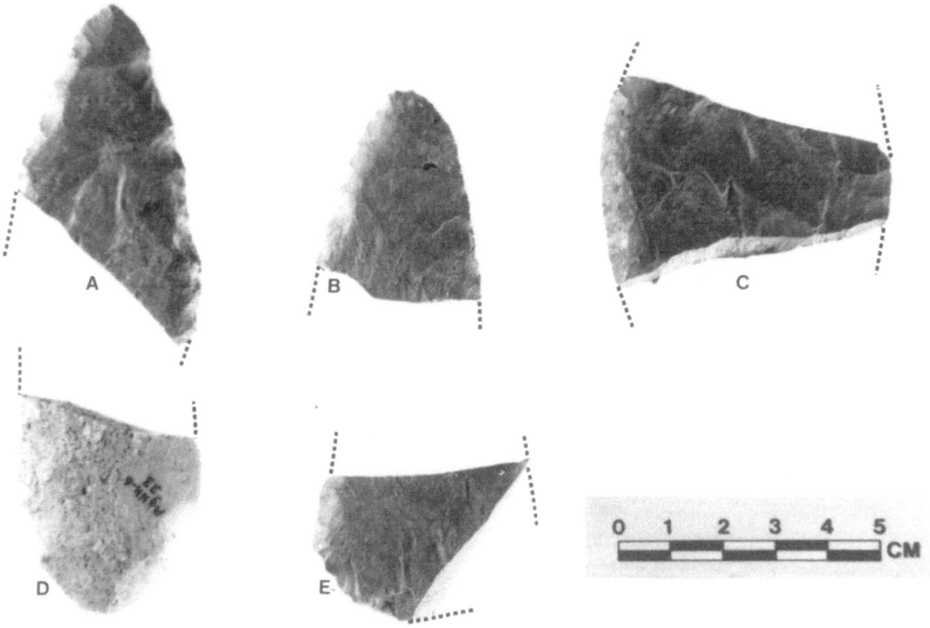

Figure 43. Alternately beveled bifaces, Thedford II site.

Figure 44. Backed biface, Thedford II site.

CHAPTER 4

Unifacial Tools

In this chapter, the unifacial tools from the site are described. Information on how particular attributes and variables of unifacial tool edges were recorded is presented in Appendix A.

End Scrapers

There are 46 tools within this general class in the Thedford II assemblage. We recognize eight types of end scrapers as well as a residual category. These types are significant in indicating differences in resharpening, hafting and, probably, use. Use differences are strongly implied by patterned variation in the size, angle and configuration of working edges and variation in tool accessories (Ellis 1984; Ellis and Deller 1988). Characteristics of the end scraper sample are summarized on Tables 22 to 30 and Figures 45 to 49.

Trianguloid End Scrapers

This type includes 10 items (Figs. 50 and 51a-b). All are complete except for one which is a lateral fragment. Seven of the tools are on flakes from biface cores, two are on side-face blanks and one is on a blank of unknown origin. The preference for large flakes from biface cores is a characteristic of this end scraper type noted at some other sites (Gramly 1982:35; Grimes et al. 1984:164).

The tools usually are roughly triangular in plan shape, have a plano-convex to trapezoid transverse section, are somewhat curved in longitudinal section and approximate the classic Paleo-Indian end scraper form. Bits are generally well centered at the distal flake end and exhibit steep, well executed, retouch. The bits tend to be relatively wide and, with one notable exception, are very shallow in relation to the rest of the end scraper sample (Fig. 46). Bit width-to-depth ratios are most often high (greater than 4:1; Fig. 47), indicating their wide, yet flat, nature. The bits also tend to be somewhat thick.

TABLE 22
Triangular End Scrapers

Variable	N	R	X	S
Length	10	24.7–58.9	36.66	10.010
Width	9	17.7–33.6	28.08	5.316
Thickness	10	5.3–12.3	7.70	1.983
Bit width	9	17.7–33.0	27.02	4.934
Bit depth	10	2.6–16.9	6.39	4.535
Bit thickness	10	3.8–12.0	6.84	2.513
Bit width/depth ratio	9	1.6–9.4	5.51	—
Curvature	7	8–12	10.28	—

N = number of observations; R = range; X = mean; S = standard deviation.

With two exceptions, the tools have continuous bilateral retouch. This retouch occurs on all the items on flakes from biface cores and on one tool on a side-face blank. The retouch serves to straighten somewhat the lateral edges and lessen the divergence of the edges from the platform. It is poorly executed.

Eight of the nine relatively complete tools exhibit tool accessories in the form of steep lateral notches or definite spurs. Six have notches on the left edge, one has a notch on the left and spur on the right and one has only a right spur. In short, there is some degree of consistency in notch/spur placement in this sample. With the exception of one tool with a notch around midpoint (Fig. 50a), all of the accessories are at or just behind the bit/lateral edge juncture. When a notch occurs just behind the bit, it isolates a thick squarish "spur" (e.g. Fig. 51b). This is not a purposeful spur or working corner. Rather, it is simply produced by gradual encroachment of the bit through resharpening. That this encroachment occurred is indicated on one specimen (Figs. 50f and 51a) where the notch is precisely at the juncture of

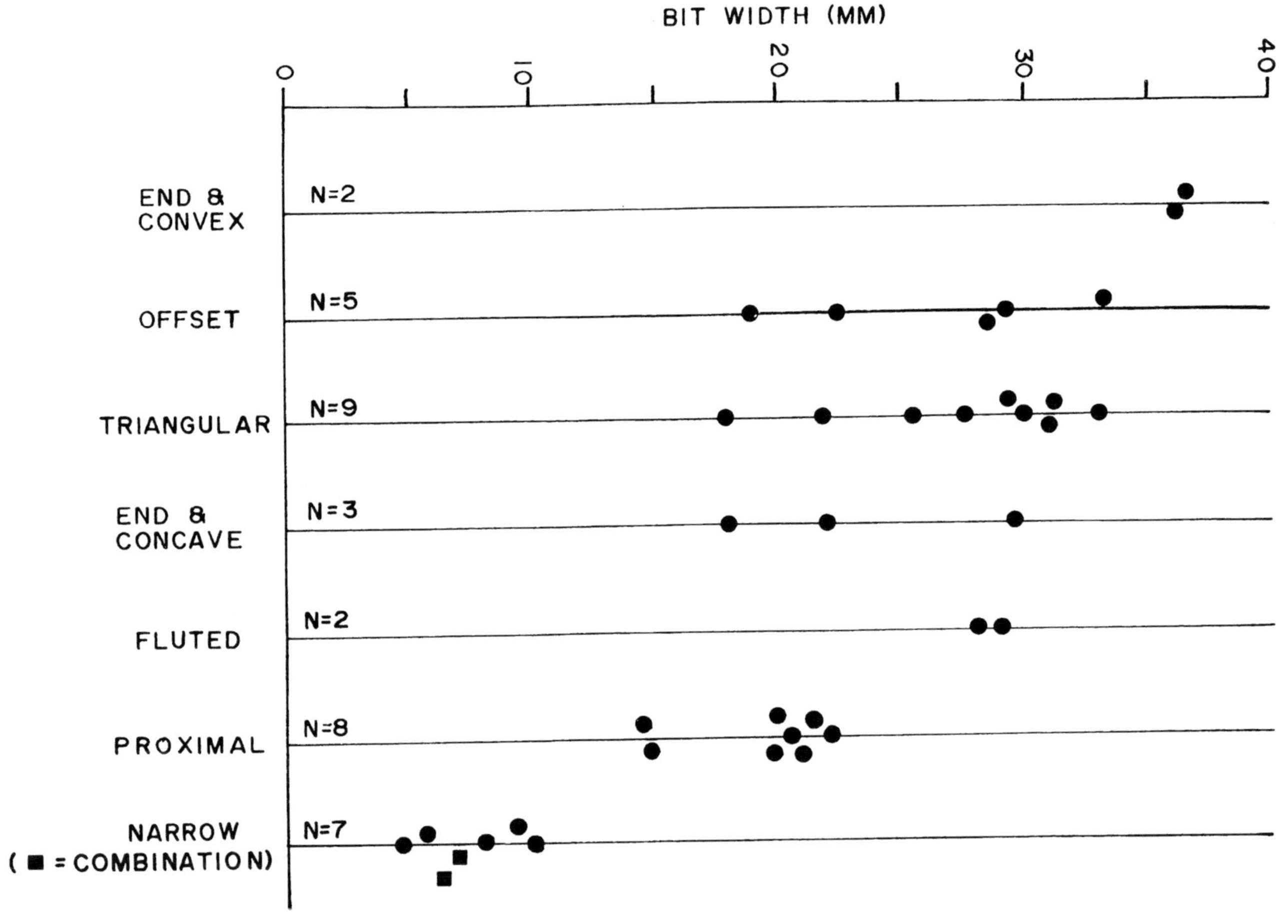

Figure 45. Bit width, end scrapers.

the bit/lateral edge having been partially encompassed by bit resharpening.

These implements, with one exception, are extensively resharpened. This is indicated by the thick, flat nature of the bits; the well executed, extensive bit retouch; the presence of well defined, abrupt junctures between the retouched bit surface and the lateral edges; and the fact that maximum thickness and width tends to occur at or just behind the bit. The exception (Fig. 50a), the longest in the sample, has the deepest bit (and hence a low bit width-to-depth ratio of 1.6:1), and exhibits the most acute-angled bit. It is the only tool without an accessory (a notch) at or near the bit. Rather, the notch is around midpoint although it is located some distance (38 mm) from the platform.

There seems little doubt that these tools were hafted. First, almost all tools have a bilateral retouch which changes outline shape and tapers the tool. This tapering would allow insertion in a socket type handle. In one case, the tapering has produced slightly inset "shoulders" on both lateral edges (Fig. 50d). In addition, we suggest that the poorly executed lateral retouch was not designed to create working edges but, rather, to simply shape outline form. It is notable that this retouch consistently occurs on flakes from biface cores but not on the flakes from other sources. The flakes from biface cores tend to markedly expand from the platform in contrast to other blank forms. Hence, some kind of lateral alteration is more often necessary in order to lessen divergence.

Second, there are attempts to thin the proximal end on some items which would again allow insertion in a socket haft. On one tool (Fig. 50d), the platform and bulb have been completely trimmed away while on another, the bulb has been partially trimmed by ventral retouch. On the one tool made on a side-face blank (Fig.

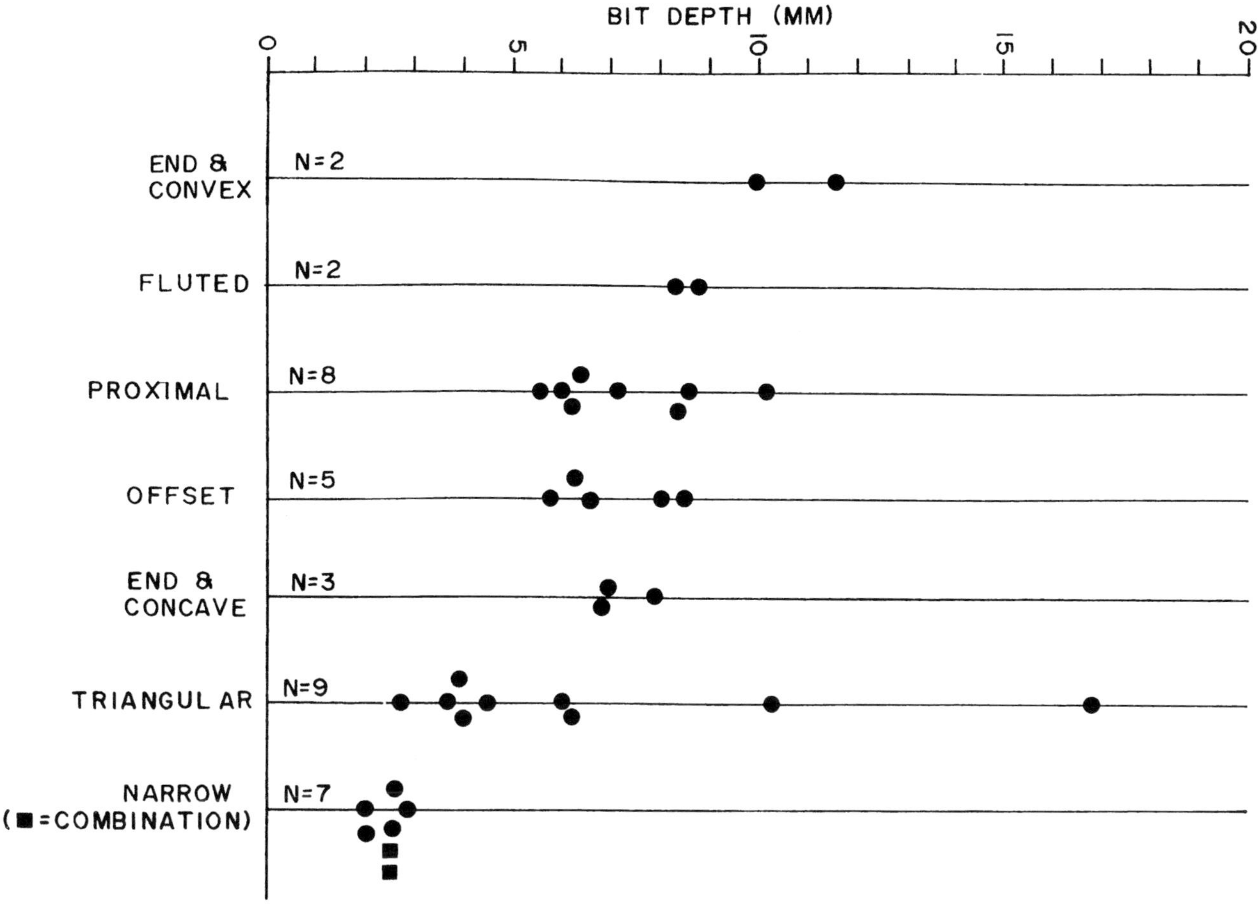

Figure 46. Bit depth, end scrapers.

50b), the lateral dorsal retouch along part or all of the lateral edges is very long and extends to or across the midline of the tool. This results in a much thinned cross-section near the proximal end. The more extensive thinning on this side-face blank, the thickest tool in the sample, suggests that flakes from biface cores were often selected for this tool type because they were thin to begin with, and needed little thinning for hafting.

Third, that the tools are highly resharpened indicates a long use-life. This resharpening is consistent with the interpretation of hafting. While stone bits for composite tools can be made quite rapidly, the time and effort required to haft such bits usually implies that they must have a more extended use-life.

Finally, all tool accessories are much removed from the proximal end. This placement is consistent with hafting in that the accessories would not be enclosed in the haft.

Asymmetrical or Offset Bit End Scrapers

There are six items in this type (Fig. 52a-f; Table 23). All are complete except for one tool which is missing part of the bit. These tools are distinctive in four respects.

First, all are made on flakes with a relatively pronounced and distal curvature such that they appear hook-shaped in longitudinal section. Not surprisingly, therefore, there is a preference for blank forms exhibiting this form of curvature, specifically, side-corner blanks. Three are on normal side-corner blanks, two are on secondary side-corner forms and one is on a side-face blank. This last blank is one of only two blanks in the Thedford II assemblage which are not from corners of the quarry block but still have pronounced distal curvature. As such, the blank's selection clearly illustrates the desire for curved bits on these tools.

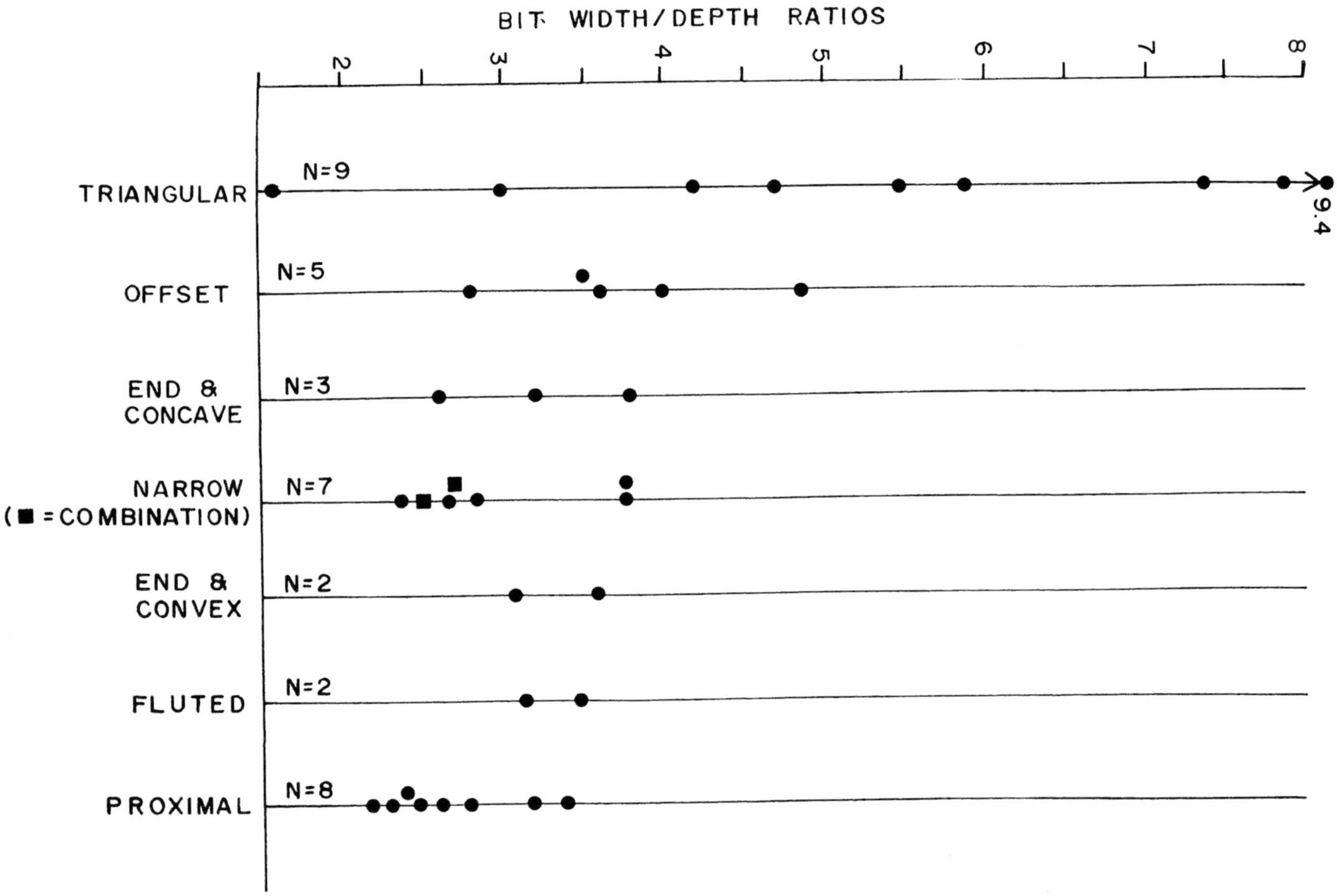

Figure 47. Bit width/depth ratios, end scrapers.

TABLE 23
Asymmetrical or Offset Bit End Scrapers

Variable	N	R	X	S
Length	6	37.5–62.8	49.13	9.769
Width	6	27.0–44.2	34.13	6.500
Thickness	6	7.0–12.2	9.83	2.264
Bit width	5	19.0–33.2	26.42	5.663
Bit depth	5	5.7–8.3	7.01	1.106
Bit thickness	6	1.7–4.7	3.36	1.121
Bit width/depth ratio	5	2.8–4.9	3.76	—
Curvature	7	4–10	6.83	—

Second, the bits are not centrally placed at the distal end. Rather, they are markedly offset or canted towards one distal end/lateral edge juncture. In some cases, all of the retouch is restricted to that corner. In every case, the bit is offset towards the right lateral edge. While it is not uncommon to recover Paleo-Indian end scrapers with offset bits, very few are as pronounced as these examples.

Third, the bits are uniformly thin (Fig. 48). Finally, bit retouch tends to be poorly executed, short and variable in length.

These tools do not appear to be highly resharpened. The thin bits relative to maximum thickness near midpoint of the tool is indicative of this, as is the larger size of most discarded items. It has been suggested that the lack of extensive resharpening on these tools was designed to maintain the distal curvature. If the tool was extensively resharpened, the curvature would be removed (Ellis 1984:196).

The lack of extensive resharpening on these tools also might suggest that the tools were not hafted. Certainly, the tools lack evidence of secondary modification suggestive of hafting as was described earlier for the trian-

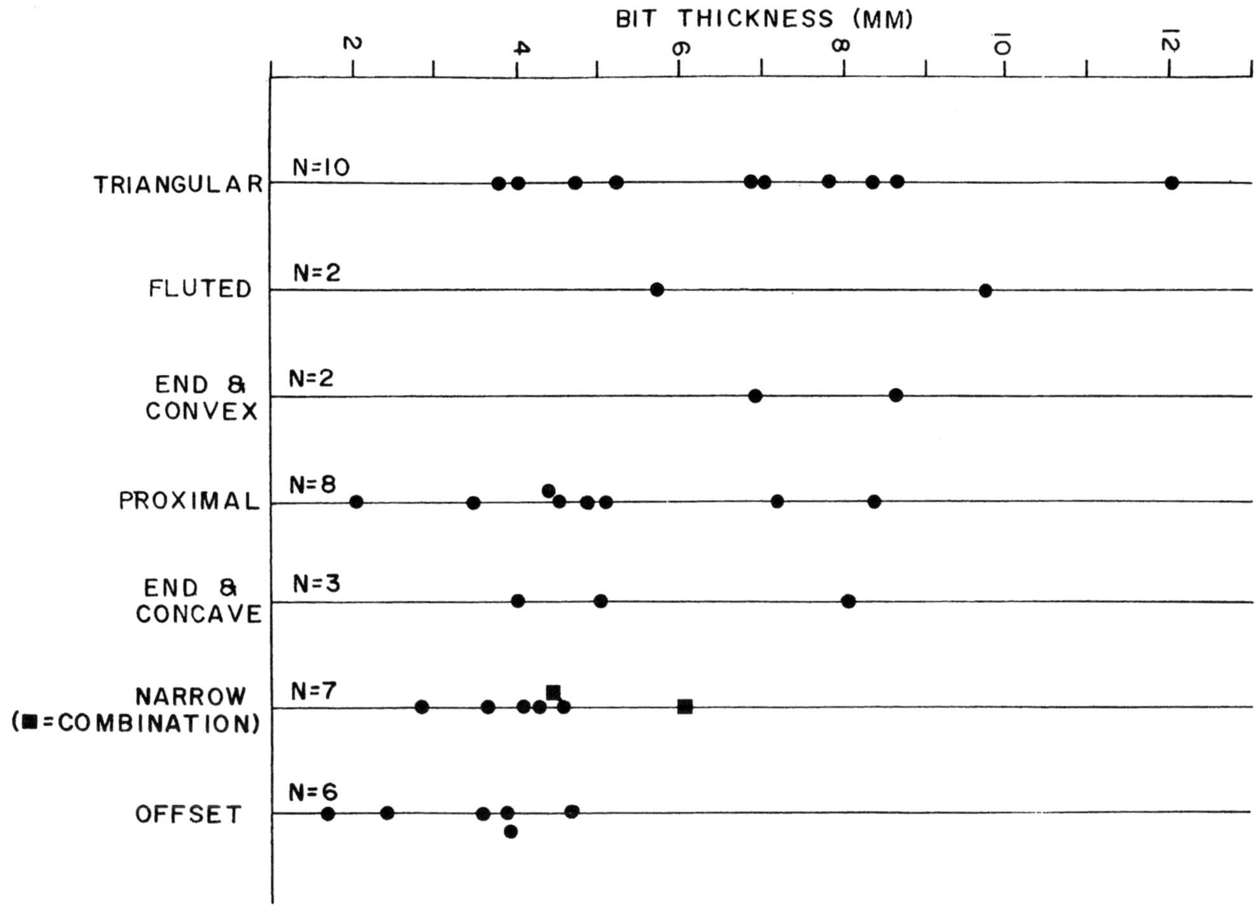

Figure 48. Bit thickness, end scrapers.

guloid forms. The major variation in proximal end shape also suggests the tools were not hafted.

Fluted End Scrapers

For lack of a better term, two implements are referred to as "fluted" end scrapers (Table 24; Fig. 52g, h). They are made on flakes with concave dorsal surfaces in transverse profile. Bits are deep (Fig. 46) and are more acutely angled than those in the other types. Also, the juncture of the bit with the lateral edges is not as abrupt as was the case with the trianguloid end scrapers. Nonetheless, since maximum thickness occurs at the distal end, they have probably been extensively resharpened.

The small size and evidence of resharpening are consistent with the idea that these tools were hafted. This inference is supported by the fact that on one the bulb and platform have been completely removed by a linear flake detachment on the ventral surface. The striking platform of the blank was used as a platform to detach this flake. A single flake representing this kind of bulb trimming (although not from this tool) was recovered in the waste flake collections (see next chapter). The removal of this flake from the scraper lent it a somewhat biconcave transverse profile. Other than for purposes of hafting, there is no viable explanation for this bulb trimming.

As these tools are wider at the proximal end than the trianguloid forms and have concave dorsal surfaces with maximum thickness along the lateral edges, it is suggested that they were hafted differently than the trianguloid forms. Perhaps they were placed in a "split-end" haft with the lateral edges exposed. This contrast with the trianguloid forms might indicate use-variability or simply, personal preference in hafting methods. One

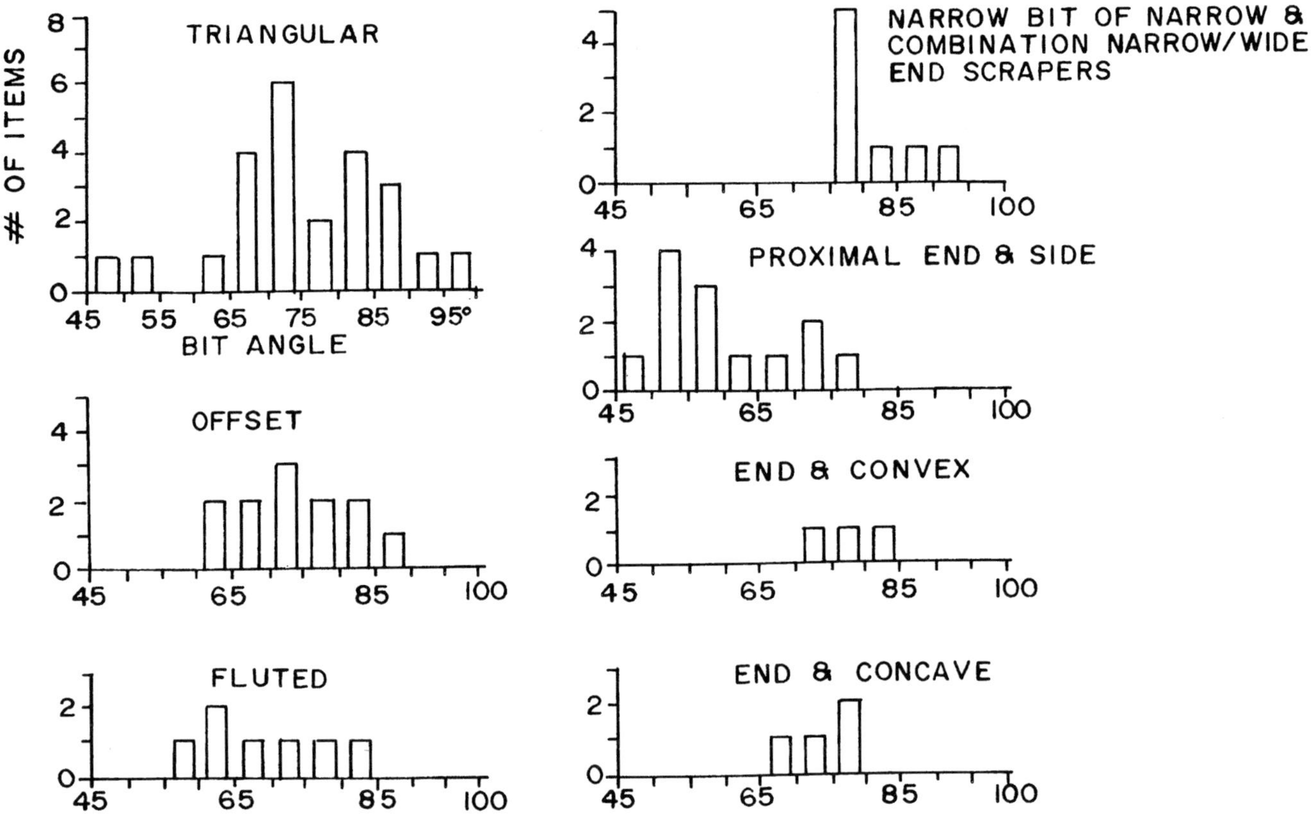

Figure 49. Bit angles, end scrapers. (Each tool is counted once for each 5° increment its angle covers; see Appendix A.)

TABLE 24
Fluted End Scrapers

Variable	FC# 232	FC# 207/245
Length	—	26.8
Width	28.1	28.9
Thickness	10.1	5.7
Bit width	28.1	28.9
Bit depth	8.9	8.3
Bit thickness	9.7	5.7
Bit width/depth ratio	3.15	3.5
Curvature	—	—

FC# = field catalog number.

TABLE 25
Narrow End Scrapers

Variable	N	R	X	S
Length	4	37.7–55.7	48.08	7.518
Width	4	20.6–32.2	27.18	4.882
Thickness	5	4.5–10.3	7.11	2.447
Bit width	5	4.8–10.0	7.54	2.355
Bit depth	5	2.0–2.8	2.38	0.363
Bit thickness	5	2.9–4.6	3.92	0.657
Bit width/depth ratio	5	2.4–3.8	3.11	—
Curvature	2	11–14	—	—

of the fluted end scrapers is on a top-face blank while the other is on a side-face blank.

Narrow End Scrapers

This type includes five tools (Fig. 53a-e; Table 25), one of which is relatively incomplete. The tools exhibit steeply retouched, narrow and shallow bits with low bit-width-to-depth ratios (Figs. 45–47) at the distal end or corner of the original blank. Most exhibit steep lateral retouch adjacent to the bit which gives the bit an isolated or "nosed" appearance in plan. This retouch probably served not only to narrow the bit but also to

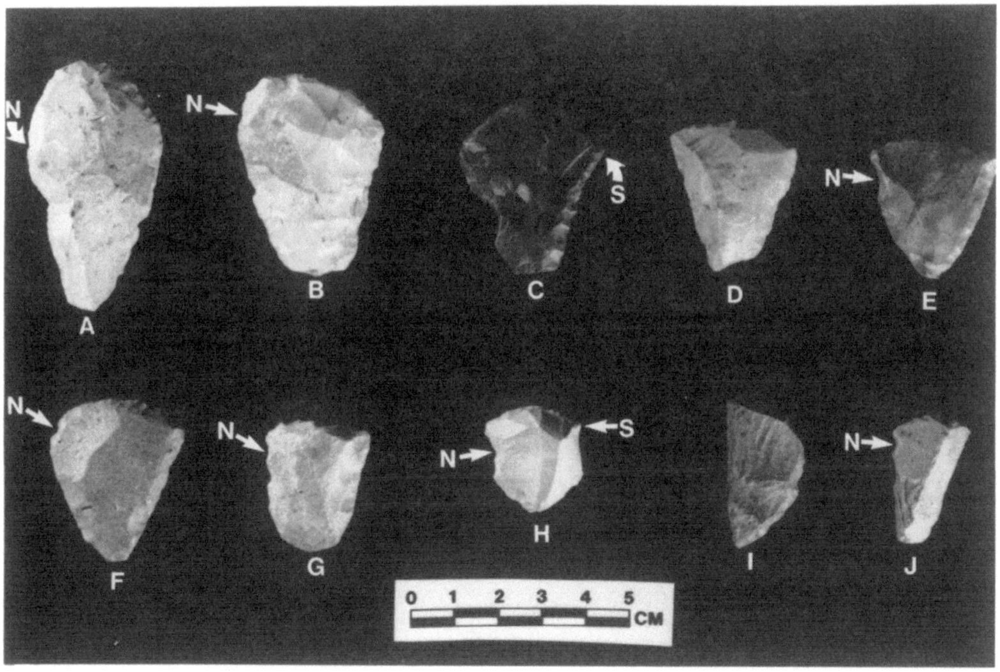

Figure 50. Triangulоid end scrapers. **N** = notch, **S** = spur.

strengthen it during use. These kinds of bits are obviously indicative of a specialized end scraper form.

Tools of this type have been referred to by a number of terms in the Paleo-Indian literature: "groovers" (Deller 1980a:67), "chisels" (Mason 1981: Plate 3:2), simply "end scrapers" (Frison and Bradley 1980: Fig. 71c-d), and "beaks" or "beaked scrapers" (Ellis and Deller 1982: Fig. 5d; Roosa and Deller 1982: 12). We suspect they have sometimes been referred to as "limaces" (cf. Gramly and Lothrop 1984:133). However, we should emphasize that they are not limaces since it is the "nosed" or narrowed end of the tool in the present sample *which is the working end*. In the case of limaces, the narrowed part *is a hafting modification* which is at the opposite end of the tool from a sometimes pointed or beaked working end. Moreover, limaces exhibit extensive retouch completely around the perimeter whereas on these tools, retouch is confined to the narrow bit area. To our knowledge, limaces do not occur at Paleo-Indian sites in southwestern Ontario.

Finally, we should note that what we are referring to as narrow end scrapers have also been called nosed end scrapers (Knudson 1973:110; Irwin and Wormington 1971:29) because those with well-centered bits can approximate the Old World type *grattoir à museau* (Bordes 1961: Plate 33–16; de Sonneville-Bordes and Perrot 1954; Tixier 1974:26). However, at least one tool in the present sample (Fig. 53e) has a bit canted towards a distal corner such that it is similar to the "lateral" nosed end scraper type (Bordes 1961: Plate 33–13; de Sonneville-Bordes and Perrot 1954), while another (Fig. 53d) with a bit placed towards one lateral edge approximates the "shouldered" end scraper type (*grattoir à épaulement;* Tixier 1974:26).

In the present sample, these varieties which might be considered separate types appear to be a product of the different kinds of blanks upon which the tools are made. On all tools, the narrow bit is flaked around a ridge on the dorsal surface, presumably to provide strength for the tool in use. Excluding the incomplete item, one is on a top-corner blank, one is on a side-face blank and two are on flakes from large biface cores. The two cases lacking well-centered bits at the distal end includes those on the biface core flakes. Such flakes generally lack well-centered ridges. Instead, the ridges tend to be off-center or to run at an angle to the blank's medial axis. As a result, if one wished to place the bit around a ridge, it would rarely be centered on blanks of this nature. This same pattern of bit placement in relation to blank form appears to occur at other Parkhill industry sites: those on biface cores have variable bit placement (e.g. Ellis 1984: Plate 8d) while those on other

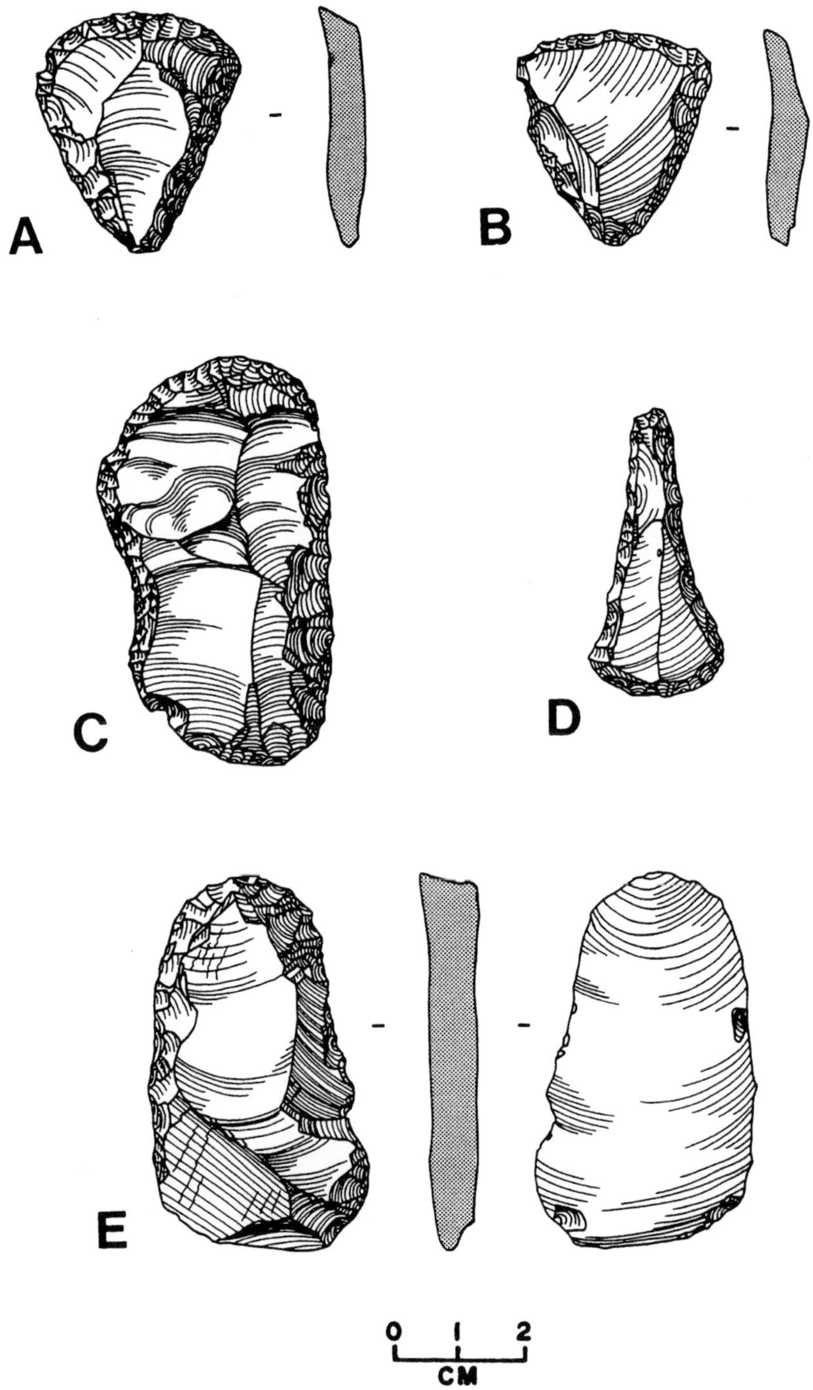

Figure 51. Various end scraper types. **a-b**: trianguloid end scrapers; **c**: end and concave side scraper; **d**: combination narrow/wide end scraper; **e**: proximal end and side scraper.

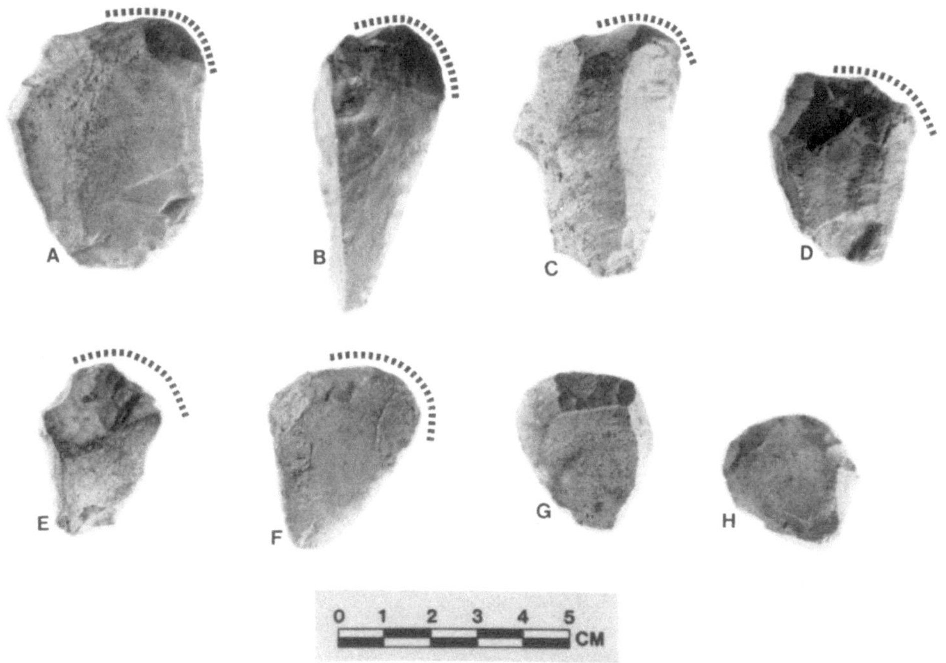

Figure 52. Offset (asymmetrical) end scrapers (**a-f**) and fluted end scrapers (**g-h**). Dashed lines show bit locations.

blank forms with well-centered ridges, such as normal top-corner forms, have well-centered bits (e.g. Mason 1981: Plate 3:2, upper row, far right). Thus, variation in bit placement can be attributed largely to the original blank selected.

There is some variation in bit width in this type (compare Fig. 53c and 53d). This variability may be a product of use-variation. However, once again, there is a possible correlation with blank type. At least two of the wider bits appear on the items made on biface core flakes while the two which can be attributed to other blank types are quite narrow. It seems possible, given that strength is a desired attribute of the bits, that the working ends on the flakes from biface cores were intentionally made wider to provide greater strength. In sum, these flakes are relatively thin compared to the other blanks. Thus, in order to compensate for their thin nature, the bits would be made wider.

TABLE 26
Combination Narrow/Wide End Scraper

Variable	FC# 513+	FC# 528	FC# 45
Length	87.6	45.5	—
Width	34.7	21.1	21.6
Thickness	12.1	7.6	5.3
Narrow bit width	6.8	6.3	—
Narrow bit depth	2.5	2.5	—
Narrow bit thickness	6.0	4.5	—
Narrow bit width/depth	2.7	2.5	—
Wide bit width	34.1	21.0	—
Wide bit depth	11.5	6.0	—
Wide bit thickness	11.3	7.6	—
Wide bit width/depth	2.96	3.5	—
Curvature	15	—	—

FC# = field catalog number

Combination Narrow/Wide End Scrapers

There are three examples of this type (Figs. 51d, 53f-h; Table 26). One is incomplete due to old breaks (Fig. 53h). These three tools incorporate a narrow bit at one end which is identical in form and size to those in the previous type (Fig. 45–49) including lateral retouch to emphasize the bit and the chipping of bits around dorsal ridges. They differ from the previous type in that they also exhibit wide, convex, steeply retouched bits at the end opposite the narrow bit. The types of flake

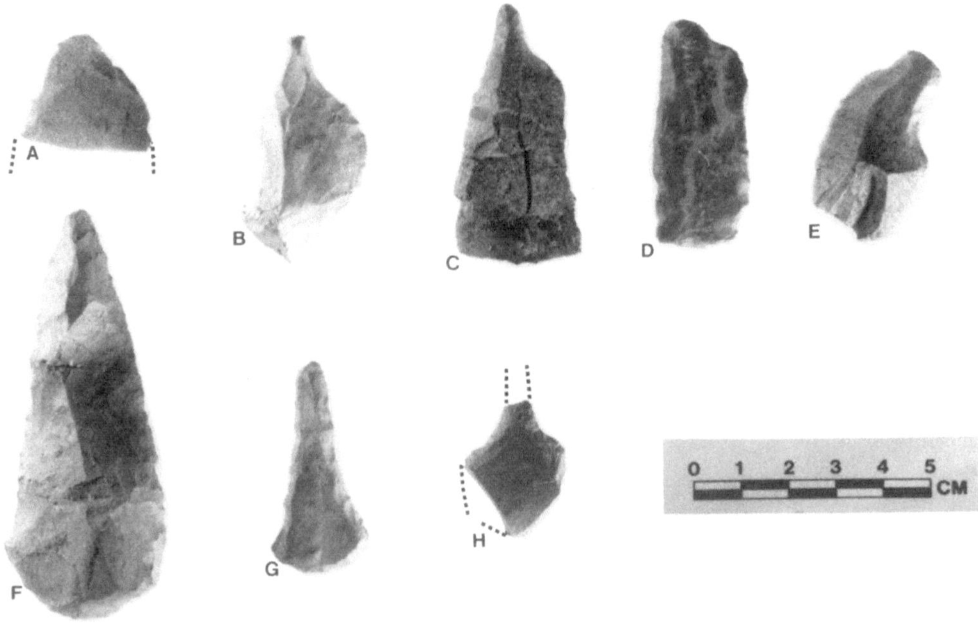

Figure 53. Narrow end scrapers (**a-e**) and combination narrow/wide end scrapers (**f-h**).

blanks on which these tools were made cannot be identified.

Proximal End and Side Scrapers

There are eight tools in this distinctive type (Figs. 51e, 54; Table 27). One tool is represented by a bit fragment while the others are complete. One tool is from a secondary top-corner blank, one a top-face blank, and the others are from unidentifiable blank forms. However, two of these latter tools which lack banding are quite large with unidirectional scars and a lack of curvature. These characteristics suggest they are from flakes approximating top-face blanks.

One defining characteristic of the type is the bits, which are often (six out of seven) at the proximal end of the blank. These bits are well centered and acutely angled (Fig. 49), and although moderately deep (Fig. 46), they are quite narrow (Fig. 48) such that they have low bit width-to-depth ratios (Fig. 47). It seems probable that the proximal ends were often selected for bit placement because they tend to be narrower than distal ends on most flake blanks.

Except for one tool with solely unifacial retouch to the left of the bit, all the tools have bilateral retouch. This retouch forms straight lateral margins, which slightly or moderately converge to the bit end. As the lateral edges generally expand from the bit, during resharpening the bits would become wider. In two cases (Fig. 54a, f), more extensive lateral retouch was applied adjacent to the bit in order to maintain a narrow width during resharpening. The result is that these two tools appear "tongued" or have inset lateral edges near the bit.

In general, retouch on lateral margins tends to be more extensive (i.e., longer, continuous, less variable in length and execution) on the edge to the left of the bit (right edge on most blanks since the bit is at the proximal end). Indeed, the edge to the right of the bit almost always exhibits only fine retouch (under 2.5 mm long) and may be largely a product of use. The left edge often (6 of 7) is heavily undercut along its length by

TABLE 27
Proximal End and Side Scrapers

Variable	N	R	X	S
Length	7	35.5–67.4	49.9	13.317
Width	8	19.6–39.2	27.8	6.180
Thickness	8	5.5–11.6	8.2	1.930
Bit Width	8	14.7–22.1	19.2	2.851
Bit Depth	8	5.6–10.1	7.3	1.586
Bit Thickness	8	2.0–8.3	4.9	1.976
Bit width/depth ratio	8	2.2–3.4	2.7	—
Curvature	6	13–15	14.3	—

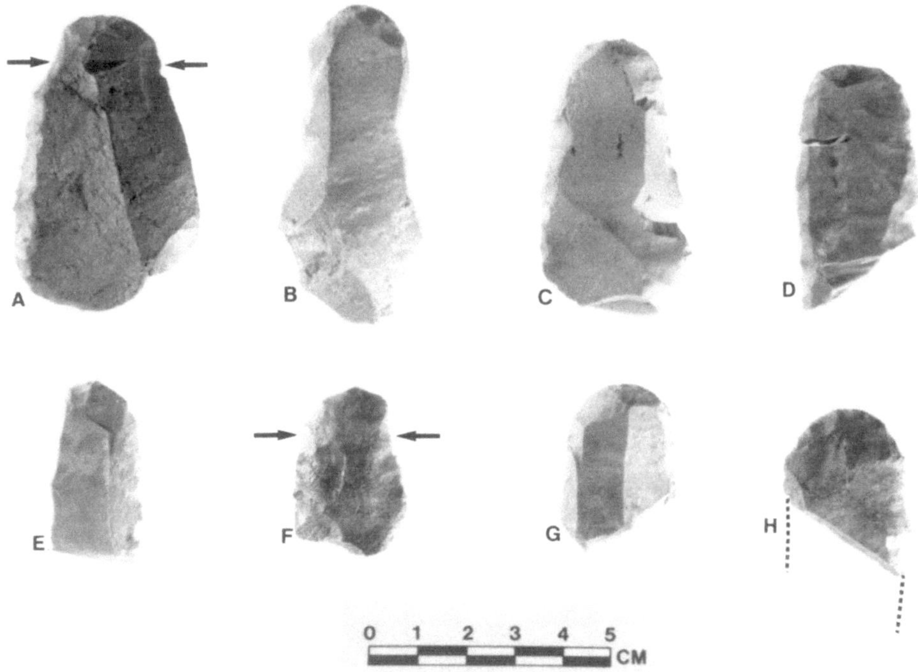

Figure 54. Proximal end and side scrapers. Arrows indicate deliberate bit narrowing.

short, step/hinge-terminated removals suggesting intensive use.

The end of the tool opposite the bit is often thick or abrupt in longitudinal profile and can (four examples) have retouch suggesting use of this end of the tool. In two cases, the abrupt end is a hinge-termination formed during the removal of the blank from the core. In one case, it is a thick remnant of a cortical surface while in another it is a "plunging" termination (Fig. 54e). On this last item, a previous removal from the same area of the core plunged through the core. This left an abrupt end on the subsequently removed flake from the same platform. Of the remaining items, one has a bend or snap break at the end opposite the bit; one has a thick, steep proximal end of the original flake; and one has probably been recycled. This last tool (Fig. 54g) has a thick steeply retouched margin upon which have been flaked two large spurs.

Two other characteristics of these tools are significant. First, the tools tend to be made on flakes which have one or two pronounced longitudinal dorsal ridges (six of nine). Second, the tools lack significant curvature in longitudinal section. This appears to be a general characteristic of most end and side scrapers as opposed to simple end scrapers.

Proximal end and side scrapers which can be unequivocably attributed to the Parkhill industry are not known from other collections we have examined. However, definite examples have been recognized by the authors in other collections from early sites in the Great Lakes region and we suspect that other examples have been illustrated in the literature (e.g., Storck 1979: Plate 27j).

End and Concave Side Scrapers

There are two tools in this type (Figs. 51c and 55a-b; Table 28). Both are made on large top-face "blade-flakes" lacking longitudinal curvature. They exhibit end scraper bits at one (Fig. 51c) or both (Fig. 55a) ends of the blank. As well, they have an extensively and steeply retouched concave lateral edge. One tool also has a straight retouched edge opposite the concave margin. The other tool exhibits a heavily worn spur adjacent to the concave edge. This combination of a spur and a concave edge frequently occurs at early Great Lakes sites (see side scraper discussion below). Tools incorporating all three working edges (the spur, concave edge and end scraper bit) also have been reported in other

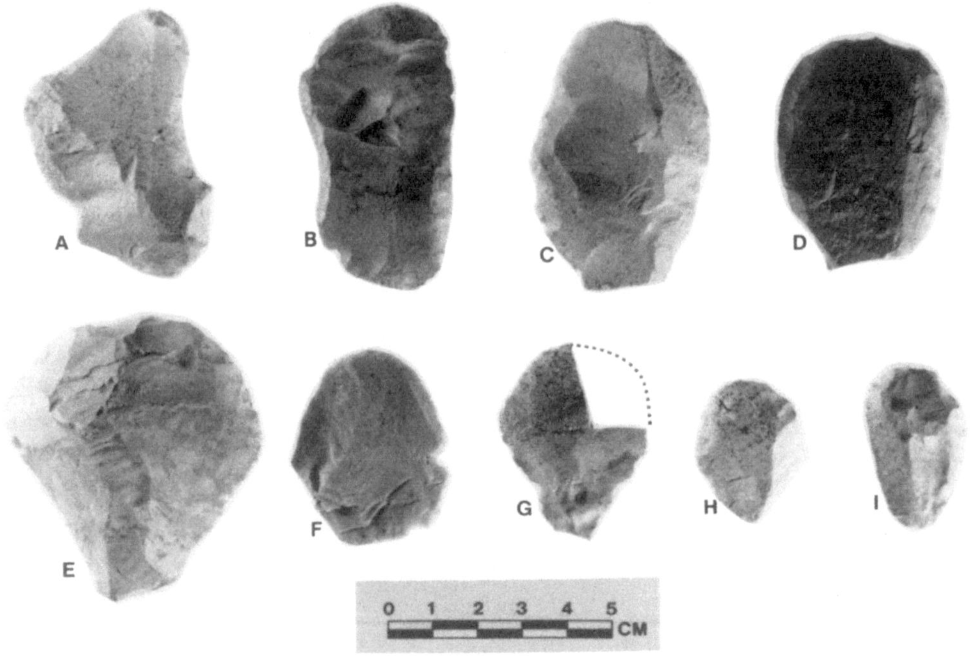

Figure 55. End scrapers, Thedford II site. **a-b**: end and concave side scrapers; **c-d**: end and convex side scrapers: **e-i**: other end scrapers.

TABLE 28
End and Concave Side Scrapers

Variable	FC# 126	FC# 19
Length	60.1	64.7
Width	40.3	35.0
Thickness	10.2	8.1
Bit width	17.9/22.0	29.8
Bit depth	6.9/6.9	7.8
Bit thickness	5.0/8.0	4.0
Bit width/depth ratio	2.6/3.2	3.8
Curvature	14	15

FC# = field catalog number.

TABLE 29
End and Convex Side Scrapers

Variable	FC# 22	FC# 324
Length	—	60.1
Width	37.7	39.5
Thickness	9.2	11.2
Bit width	36.3	36.1
Bit depth	10.0	11.6
Bit thickness	6.9	8.6
Bit width/depth ratio	3.6	3.1
Curvature	15	13

FC# = field catalog number

early assemblages from the Great Lakes area (Ellis and Deller 1982: Fig. 41).

End and Convex Side Scrapers

Two relatively complete tools are of this type (Table 28). They exhibit well-centered bits at the distal end as well as extensively retouched, steep (55–90°), convex edges on one (Fig. 55c) or both (Fig. 55d) lateral margins. On one tool, the lateral margins are heavily undercut suggesting they are use margins. Combined with their thick proximal ends, this might suggest the tools were not hafted. Yet, at the nearby Parkhill site a number of these items were present with snapped bits. The pressure needed to snap these thick blanks raises the possibility they were hafted. One of the tools (Fig. 55c) is on a side-face blank while the other, although of Bayport chert, may be from a top-face blank or equivalent thereof based on its size, lack of curvature and the presence of large unidirectional dorsal scars.

Other End Scrapers

There are nine other end scrapers not including one tool recycled into another tool form (backed and

TABLE 30
Other End Scrapers

Characteristic	Flake End Scraper	Large End Scraper	Small Oval End Scraper	Small Oval End Scraper	Other End Scraper
Figure #	55g	55e	55i	55h	55f
Length	45.1	63.5	31.4	31.4	43.2
Width	32.1	56.7	21.5	23.2	35.4
Thickness	5.8	17.3	7.1	8.6	8.2
Bit width	—	56.7	16.7	21.2	19.5
Bit depth	—	25.4	7.8	8.0	4.3
Bit thickness	—	11.0	4.6	7.5	2.9
Bit width/depth ratio	—	2.23	2.14	2.65	4.5
Bit angle	45–50	40–60	70–75	55–80	60–65
Curvature	15	10	—	—	6

snapped uniface; see below). Four of these tools are represented simply by bit ends. One of these was made on a secondary top-corner flake. Another bit might be from a trianguloid end scraper. Still another bit is quite large and is made on a top-struck flake blank. This fragment may be from an end and convex side scraper as described above or it may be from a tool similar to a very large end scraper to be described below.

The remaining five tools (Table 30) include a thin end scraper with a fine (under 2.5 mm), acute-angled (40–50°) bit (Fig. 55g); a tool on a side-face blank (Fig. 55f) with a narrow (19.5 mm), shallow (4.3 mm), thin (3 mm) curved bit; and a tool on a large top-face blank (Fig. 55e), with a wide (56.7), deep (25.4) bit and some ventral retouch. The platform has been trimmed away indicating the tool was probably hafted.

The final two tools are small forms with roughly oval outlines (Fig. 55h, i). One is from a flake from a biface core while the other is from a blank of unknown origin. The former has a slightly denticulated bit suggesting recycling while the latter exhibits ventral bit retouch, probably from tool edge rejuvenation.

Side Scrapers

There are 15 tools in this class which have continuous, extensive (i.e., extending more than 2.5 mm back from the edge) retouched edges along one or both lateral margins. With a few possible exceptions noted below, there is little patterning within the class. This could be the result of constraints in tool production—perhaps the only constraint was that the tool could be held comfortably in the hand. Moreover, it is probable that morphology and even the presence of retouched edges are in many cases a product of edge rejuvenation rather than manufacture *per se*. As a result, only a simple morphological classification of these tools was carried out based on the shape in plan view of working edges and whether or not retouch occurs on one or both lateral margins. Characteristics of each individual tool are given on Table 31.

The assemblage includes 11 "single-edged" tools. These consist of four single convex forms (Fig. 56a-d), one of which has intermittent retouch opposite the scraper margin and might be considered a "backed knife" (*couteau à dos*); fragments of probably two single straight forms (Fig. 56e); a single tool (Fig. 56f) with a concavo-convex margin; and four tools with single concave margins (Fig. 56g-i).

One of the single convex forms (Fig. 56a) is on a large top-face blade-flake and exhibits a bend or snap break with evidence of utilization on the margin opposite the convex edge. At least three tools with the same characteristics (manufacture on a top-face blade-flake; presence of a single, well-retouched convex margin; and presence of a bend-break or right-angled edge—often with evidence of use) are known from other Parkhill industry sites in the area (see Ellis 1984:219). This suggests they may be a specialized tool form, designed for use in specific task contexts.

On one of the concave tools (Fig. 56h), the lateral retouch extends around the corner of the flake onto its distal end. Another one of these concave scrapers or "spokeshaves" exhibits a spur at each end of the working edge (Fig. 56g). One of these spurs is at the juncture of the concave edge and a snap across the proximal end and might be fortuitous (there is no clear evidence of use). However, the other spur has retouch on that side of the spur opposite the concave edge and is undoubtedly intentional. In our experience, and as implied in the earlier discussion of the end and concave side scrap-

TABLE 31—Side Scrapers

Type	Blank Type	Length	Width	Thickness	FC#
Single convex	flake from biface core	48.5	36.3	4.9	682+
Single convex	top face	59.4	48.7	9.3	68+
Single convex	normal side corner	—	31.3	7.9	155+
Single convex	flake from biface core	55.3	—	5.4	75+
Single straight	unknown	—	—	—	215
Single straight	side face	41.4	24.9	7.9	555
Single concavo-convex	unknown	45.4	33.1	10.4	474
Single concave with spur	side face	44.0	29.5	5.5	681
Single concave	flake from biface core	39.4	35.2	6.2	242 (not illus.)
Single concave	flake from biface core	36.4	35.8	8.5	329
Single concave	normal side corner	41.5	24.7	11.0	170+
Double convex	top face	46.6	24.3	7.4	367 (not illus.)
Double convex	unknown	—	30.0	5.7	254
Double convex	top face	—	25.1	5.1	275
Double alternate	unknown	52.8	19.9	4.6	76+

ers, spurs in the Great Lakes area are frequently associated with tools exhibiting a concave margin, particularly side scrapers. For example, at the Crowfield site (Deller and Ellis 1984:47), the only two side scrapers with spurs are forms also incorporating a concave margin. Similarly, at the Parkhill site, the only four side scrapers with spurs are concave ones (e.g. Deller 1980a: Plate K2). Other examples of this association have been illustrated in the Great Lakes literature (e.g. Fitting et al 1966: Fig. 19x; Storck 1978: Fig. 5c, bottom row, 1979: Plate 7i; and perhaps Ritchie 1980: Plate 10–6).

The remaining tools are double-edged. They include three somewhat fragmentary double convex forms (Figs. 56j-k) and one double alternate form (Figs. 56l, 57c) with unifacial retouch on one face along one edge and the opposite face on the other edge.

Combination Notch/Borers/Denticulates

The seven tools in this class incorporate notches, thick massive spurs or borers, and retouched (usually denticulated) working edges in various combinations (Table 32; Fig. 58). One is notable in that it may have been recycled (Fig. 58f). This tool is on a secondary side-corner struck flake. It has a thick "bit" towards the right corner of the distal end which has remnants of five, closely spaced, thick borers. All have been heavily undercut and damaged by use. Both lateral margins of this tool have a very fine denticulated working edge. Given the offset right bit and the tool's manufacture on a side-corner blank with pronounced distal curvature, it seems possible that it was originally an offset end scraper.

This general class of tool consistently occurs on early sites in the Great Lakes region. The Parkhill site collections contain some examples and at least two are known from Hi-Lo contexts (Ellis and Deller 1982: Fig. 4n, o). Undoubtedly they represent expediently used tools employed in a wide range of tasks.

Bend-Break Tools

This class includes three items (Fig. 59g-i). As with some tools placed into other types or classes, these tools incorporate a bend or snap break with evidence of use

LEFT LATERAL EDGE				RIGHT LATERAL EDGE			
Edge Form	Edge Thickness	Edge Angle	Retouch Type	Edge Form	Edge Thickness	Edge Angle	Retouch Type
—	—	—	—	CO	2.2	35–45	N
—	—	—	—	CO	4.0	45–90	N
—	—	—	—	CO	4.6	65–75	N
CO	4.2	50–65	N	—	—	—	—
—	—	—	—	ST	3.2	40–55	N
—	—	—	—	ST	5.2	50–75	N
CC/CO	4.4	45–55	N	—	—	—	—
—	—	—	—	CC	2.7	40–65	N
CC	4.0	65–75	N	—	—	—	—
—	—	—	—	CC	3.0	55–65	N
—	—	—	—	CC	4.2	60–65	N
CO	4.4	65–75	N	CO	2.3	70–75	N
CO	1.4	60–75	N	CO	3.2	40–65	N
CO	2.9	40–80	N	CO	2.7	40–55	N
CO	3.2	55–65	N	CO	4.5	55–70	I

CO = convex; CC = concave; CC/CO = concavo-convex; ST = straight; N = normal retouch; I = inverse retouch.

along one margin. Such tools have been noted in a number of Paleo-Indian collections since they were first reported from Folsom contexts by Frison and Bradley (1980:91–97). It is possible that they are Paleo-Indian equivalents of burins.

Other Unifacial Tools

Backed and Snapped Unifaces

This distinctive tool form is represented by six examples (Figs. 57d, 59a-f; Table 33). Tools of this type also have been reported from Parkhill industry contexts at the McLeod site (Ellis 1984:223; Ellis and Deller 1988). Two varieties are recognized in the Thedford II sample.

The first variety includes five items (Fig. 59a-e). All are roughly rectangular in plan view and are wedge-shaped or offset triangulars in transverse section. All of these tools are made of side-corner removed flakes—two being normal and three being secondary in derivation. The narrow dorsal facet of the wedge or offset triangle in section is the old top surface of the core. It forms the "back" of the tool and is usually (four of five) a cortical surface lacking retouch. The lateral edge opposite the back (i.e., the thin edge of the wedge in transverse section) exhibits a short (<3.9 mm), continuous, normal or inverse retouch. In addition, one or both ends of the tool exhibit snaps or bend-breaks at right angle, in plan, to the retouched edge and "back." In all cases, the lateral retouch on the thin edge extends around the corner onto the bend-break surface. This retouch can extend entirely across the snapped edge or just around the corner from the retouched lateral edge. The latter suggests the thick corner juncture of the snap and retouched edge was important in the tool's use.

Three of these tools are notable in that there is evidence they were made on the site. Two fit together (Fig. 59a, b) indicating that they were made on the same blank snapped across its midsection. Subsequently, retouch was applied to the breaks and to the thin lateral edges on each item. The third (Fig. 59c) fits together with an end scraper bit which was snapped off either purposefully or during use. After breakage, the bit was discarded and retouch was applied across the break on the proximal half of the flake. The lateral retouch also appears to have been applied subsequent to the break.

The second variety of this tool type is represented by one item (Fig. 59f) on a normal side-corner blank. This tool exhibits the same combination of edge forms

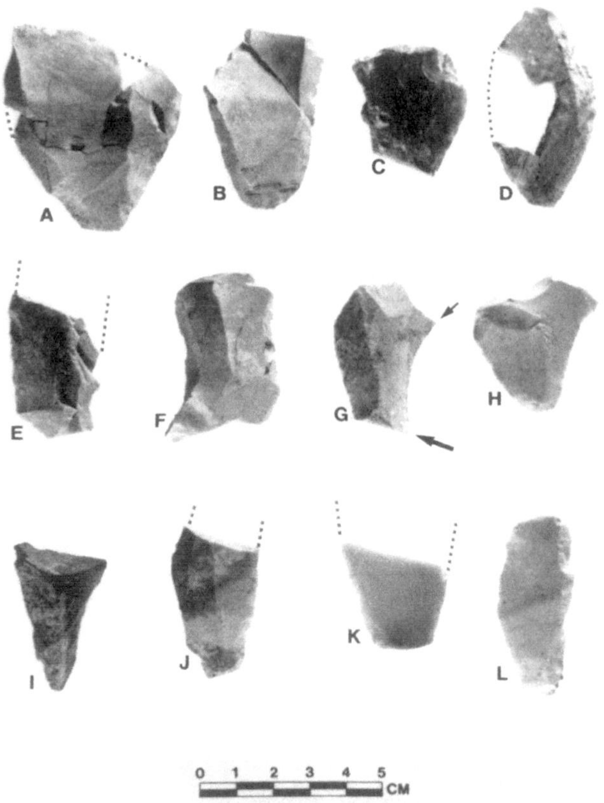

Figure 56. Side scrapers. Arrows show spur locations.

TABLE 32
Combination Notch/Borers/Denticulates

FC#	Figure	Length	Width	Thickness
491	58a	50.1	37.2	6.1
668	58b	64.8	46.7	9.1
554	58c	54.5	45.4	16.1
92 +	58d	49.7	35.7	10.4
279	58e	49.6	37.4	7.1
603	58f	63.2	36.3	11.5
666	58g	20.9	20.4	7.0

FC# = field catalog number.

as the previous items (acute retouched edge opposite a back; retouched snap along an edge at right angle, in plan, to the acute and back edge). The difference is that these edge forms occur on different margins. The back is a snap at the proximal end while the opposite acute retouched edge is at the distal end. The latter was formed by a snap which encountered a natural fracture plane in the material and as a result, produced an acute rather than a right-angled edge in profile. The snapped retouched edge is on a lateral margin.

Piercers or Spurs

Eighteen simple piercers or spurs ("gravers") were recovered (Table 34; Figs. 57a-b, 60). Sixteen of these are relatively complete while two are fragmentary. It is possible that the latter two items are fragments of composite tools such as the side scrapers with spurs noted earlier. However, they appear to be from much smaller, thinner, flakes and the spurs are more diminutive.

One piercer is made on a top-face blank, three are made on flakes from biface cores and two are made on side-face blanks. In addition, two are placed on end biface thinning flakes (e.g. Fig. 57a) and another is made on a normal biface thinning flake identical to those flakes recovered unmodified in the flaking debris from the site (see next chapter). All of the remaining piercers are of blanks of unknown type.

A minimum length of 1 mm was imposed to define a piercer. On this basis, 43 spurs are represented in the assemblage. On complete items, from one to nine spurs are present with two spurs being the most common (Table 35). While one spur is simply a natural pointed projection, the others are variously retouched into shape. The retouch applied is steep and probably served to strengthen the tool in use as well as to form the working edge. As at Debert (MacDonald 1968:100), another means used to provide strength for spurs was to flake them around a dorsal ridge on the blank. However, this occurs on only 14 (32.5%) of the 43 spurs. Another common method used to strengthen the spur is to place it at the corner of one (thirteen examples) or two (one example) right-angled edges. In these cases, the juncture of the right-angled edge and the flake surface forms a strengthening ridge. The right-angled edges can be formed by snap breaks (thirteen examples or edges), natural fracture planes (one) or weathered surfaces (two). This right-angled spur margin can be either retouched or left unmodified.

Several kinds of spurs can be distinguished based on the nature of the retouch. These different spur types can co-occur on the same tool. Thirty were formed by a bilateral retouch on the dorsal surface (two edge normal-unifacial), one was created by retouch from both margins on both faces (bifacial) and one was produced by bilateral retouch on the ventral flake surface (two edge inverse-unifacial). The remainder were formed by alternate retouch (three), one edge normal-unifacial re-

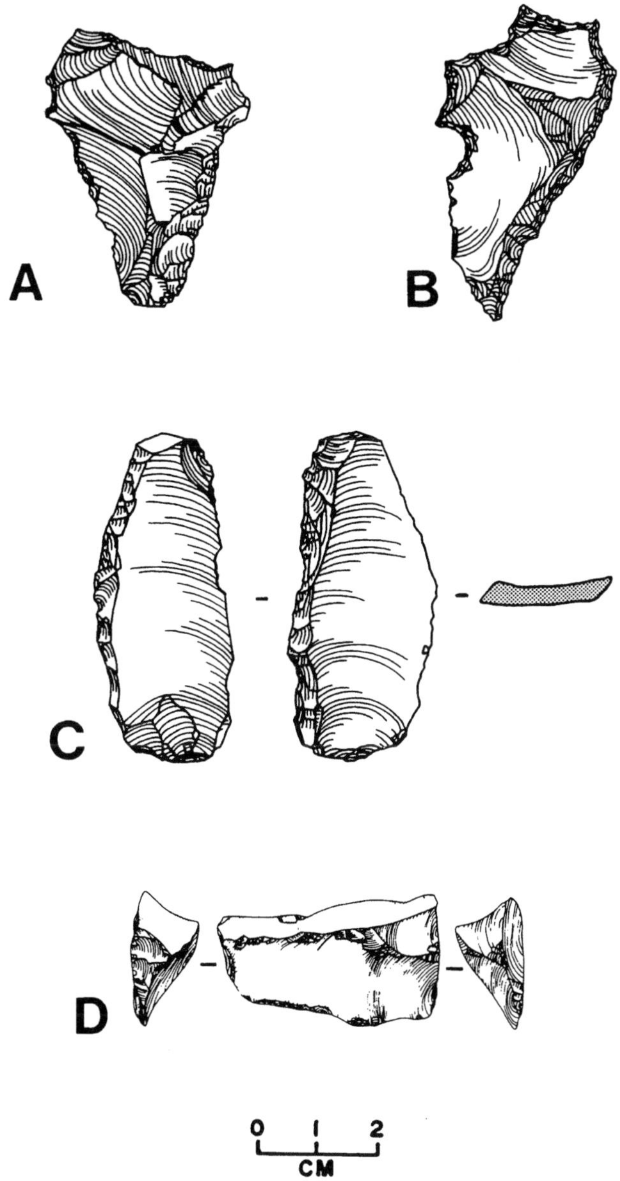

Figure 57. Miscellaneous unifaces. **a-b**: multiple piercers; **c**: double alternate side-scraper; **d**: backed and snapped uniface.

touch (six) or one edge inverse-unifacial retouch (one). In these last seven cases, the opposite margin is an unaltered right-angled edge.

At least one of these tools was definitely made on the site. A small side-face blank was snapped in half and both halves were recovered (Fig. 60o). The proximal flake end was simply discarded. On the distal half, two notches were chipped just behind the snap to isolate spurs at both corners of the break. In one case, the retouch was also applied to the break surface.

Denticulates

Three tools have simply denticulated edges on one or more margins (Fig. 61a-c). One (Fig. 61a) is notable in that it has marginal ventral modification as well as

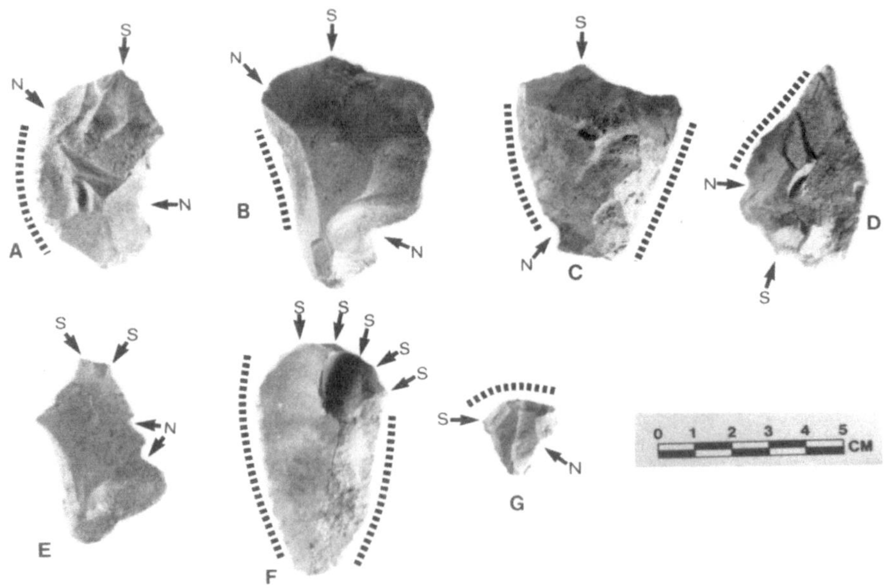

Figure 58. Combination notch/borers/denticulates. **S** = spur, **N** = notch, dashed lines show location of retouched or denticulated edges.

TABLE 33
Backed and Snapped Unifaces

Variety #1

Variable	N	R	X	S
Length	5	31.3–36.0	33.46	2.351
Width	5	21.7–26.7	24.84	2.341
Thickness	5	7.2–11.4	9.69	1.925

Variety #2

Length = 25.3	Width = 24.4	Thickness = 8.2

thinning flake scars over the complete dorsal surface. This perhaps indicates attempts to recycle a unifacial tool into a biface prior to its being denticulated.

Retouched Flakes

Six tools exhibit a fine continuous to intermittent retouch along one or more edges (Fig. 61d-i; Table 36). Some of these approximate the "thin flake knives" reported from several Paleo-Indian sites while others could be easily classified as "raclettes" (cf. Bordes 1972: Fig. 5; Frison and Stanford 1982:50).

Other Tools

This is a residual category that includes seven items (Table 37). One is a large "blocky" scraper made on a core fragment rather than a flake. It exhibits long, steep, thick, continuous retouch along one side and end (Fig. 62a). A second tool (Fig. 62b) is a transverse convex scraper on a blank of unknown origin. Unlike the vast majority of the tools on Collingwood chert, this tool exhibits curved banding in plan view. A third tool (Fig. 62c), on a secondary side-corner flake, exhibits a short, continuous, steep retouch along one straight lateral and one straight distal edge. This tool also exhibits a spur to the left of the platform.

Another tool is a beak made on a top-face blank fragment (Fig. 62e). At one end is a very large, thick, pointed projection formed at the juncture of a lateral edge and a break across the end. This projection was subsequently emphasized by steep marginal retouch. The tool also exhibits a convex, acute-angled scraping edge along one margin. Beaks are a common find on Great Lakes Paleo-Indian sites (e.g. Storck 1979:40).

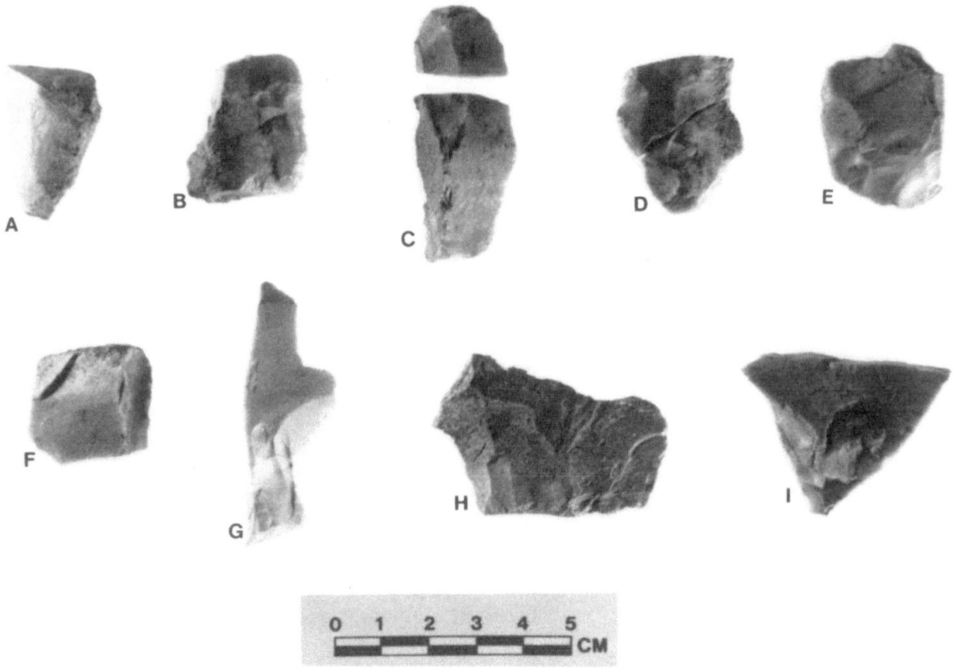

Figure 59. Backed and snapped unifaces (**a-f**) and bend-break tools (**g-i**).

TABLE 34
Piercers

Variable	N	R	X	S
Length	13	16.1–48.7	35.22	9.313
Width	16	16.1–50.4	27.23	9.133
Thickness	14	2.5–8.4	4.79	1.724
Spur length	36	1.0–4.2	2.27	0.806
Spur width	38	0.8–2.8	1.63	0.474
Spur thickness	41	0.9–2.8	1.60	0.444

One tool (Fig. 62d) is on a top-face blank. At one end is a thick spur which is "squarish" in plan and resembles those items referred to as "chisel gravers" by F. Roberts (1935). Similar tools occur at Parkhill industry sites such as McLeod (Ellis 1984:22).

Of the remaining tools, one is a simple notch on a small flake of unknown origin (Fig. 62g). The final tool is on a flake from a biface core (Fig. 62f). Along the right edge are two, slightly concave, working edge segments. At both ends of the edge and in the middle separating the two concave sections are narrow, slightly convex projections.

Tool Fragments

The remainder of the tool assemblage consists of 85 tool fragments formed by heat fractures, plow breakage and probable use breaks. Most of these fragments appear to be from side scrapers or end scrapers and one was from a composite tool with a spur. Six of the fragments are of Bayport chert and thus may be associated with later occupations on the site. It should also be noted that three biface thinning flakes, one channel flake and a large subgreywacke flake included in the waste flakes (see next chapter) may have been used.

Summary

The preceding analyses have recognized eight general classes of tools. Within these classes it has been possible to isolate 22 tool types if we include some tool forms such as backed bifaces which, although rare in the present assemblage, are known to occur on other Paleo-Indian sites in Ontario. The distribution of the various tool types and classes by the blank types is given on Table 38.

As suggested earlier, and argued in detail elsewhere (Ellis and Deller 1988), the highly distinctive nature of

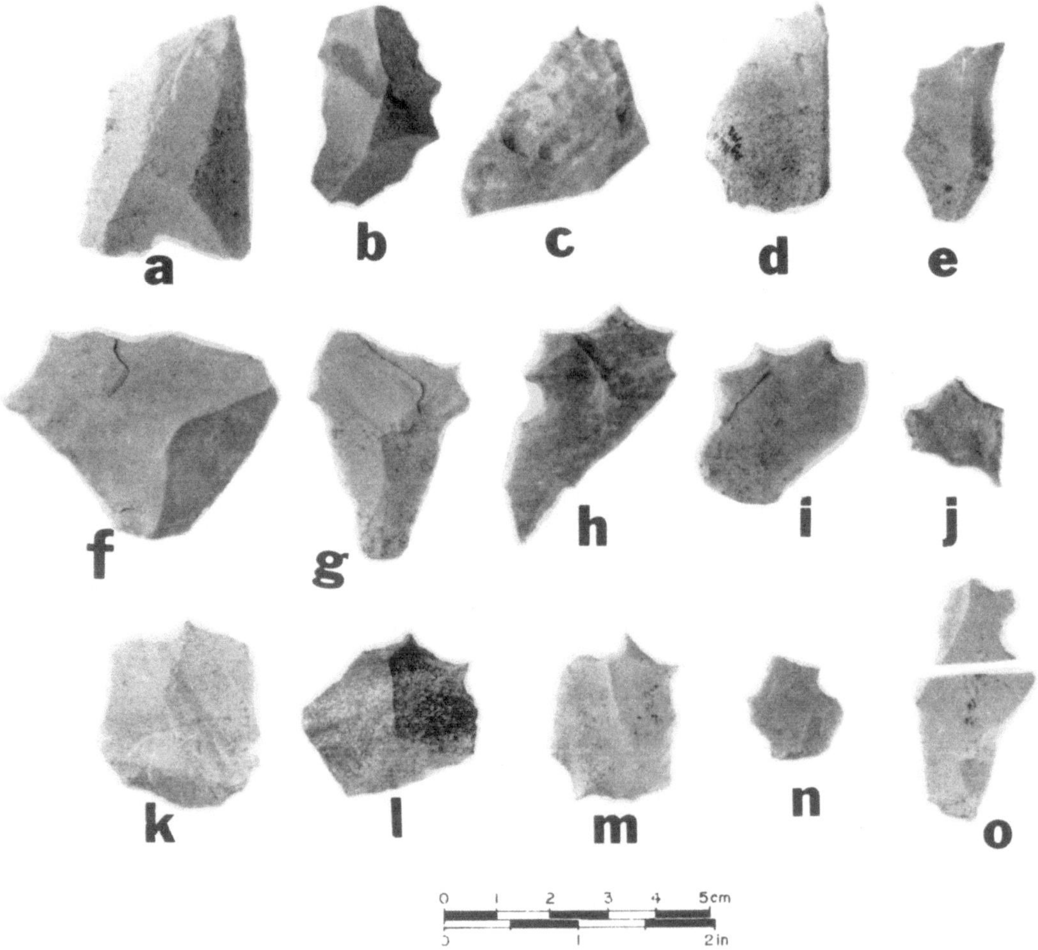

Figure 60. Piercers or gravers. Note that **m** is from the McLeod Site.

TABLE 35
Piercer Spur Counts

Number of Spurs	Number of Tools
1	4
2	6
3	4
4	1
9	1

some of the tool types and by definition, their recurrent appearance, suggests they are significant in terms of use-variation in the assemblage. Although there may be some use-variability within such types, this is thought to be minimal in comparison to some of the tool categories often used to characterize inter-assemblage activity variation such as end scrapers. Distinctive types of use-significance include: fluted points, alternately beveled bifaces, backed bifaces, channel flake points (although these may be simply "toys" as Roosa [1977b:115] has suggested), trianguloid end scrapers, offset end scrapers, narrow end scrapers, combination narrow/wide end scrapers, proximal end and side scrapers, backed and snapped unifaces, piercers or gravers

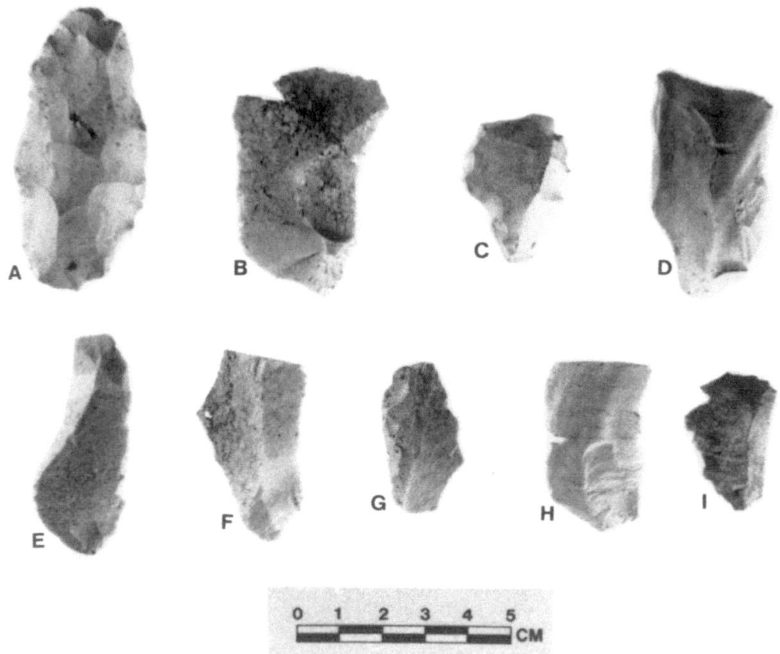

Figure 61. Retouched flakes (**d-i**) and denticulates (**a-c**).

TABLE 36
Retouched Flakes

Variable	N	R	X	S
Length	6	32.7–54.4	42.75	8.758
Width	6	18.0–29.2	23.85	3.735
Thickness	6	1.1–7.7	4.30	2.170

TABLE 37
Other Unifacial Tools

Type	Figure	Length	Width	Thickness
"Blocky" scraper	62a	62.8	57.4	16.2
Transverse convex scraper	62b	37.2	45.8	13.9
Scraper/graver	62c	39.7	36.5	10.8
Beak	62e	36.4	44.7	10.1
"Squared" beak	62d	34.0	24.3	9.0
Notch	62g	24.2	13.9	7.3
Concave-convex scraper	62f	54.8	35.0	7.1

and, perhaps, concave side scrapers with spurs, large single convex side scrapers with bend-breaks, end and concave side scrapers, and beaks.

Since the Thedford II assemblage is small, we do not believe it includes all of the use-significant tools associated with the Parkhill industry in Ontario or for that matter, with Paleo-Indian sites in general in the Great Lakes region. It is entirely possible that with the examination of other tool samples, tools represented in the present assemblage by single examples may be found to occur frequently on Paleo-Indian sites. The single "backed biface" in the assemblage which is a tool type known to occur at several other sites is a good example. Moreover, it is possible that there are some tool forms characteristic of these industries which are not represented at all in the Thedford II assemblage. For instance, one tool recovered from the Parkhill site is unifacial with a pointed end and marginal bifacial retouch. A similar tool has been illustrated from the Gainey site (Simons et al. 1984: Fig.9b) and we know of other examples from additional sites. These tools are very similar if not identical to those referred to as "Mungai knives" at Meadowcroft Rockshelter (Adovasio et al. 1983: Figs. 13.7a, 13.9a). Another example: the Thedford II assemblage lacks distinctive, elongated items with wide stems and narrow, sharp, working tips which we have called

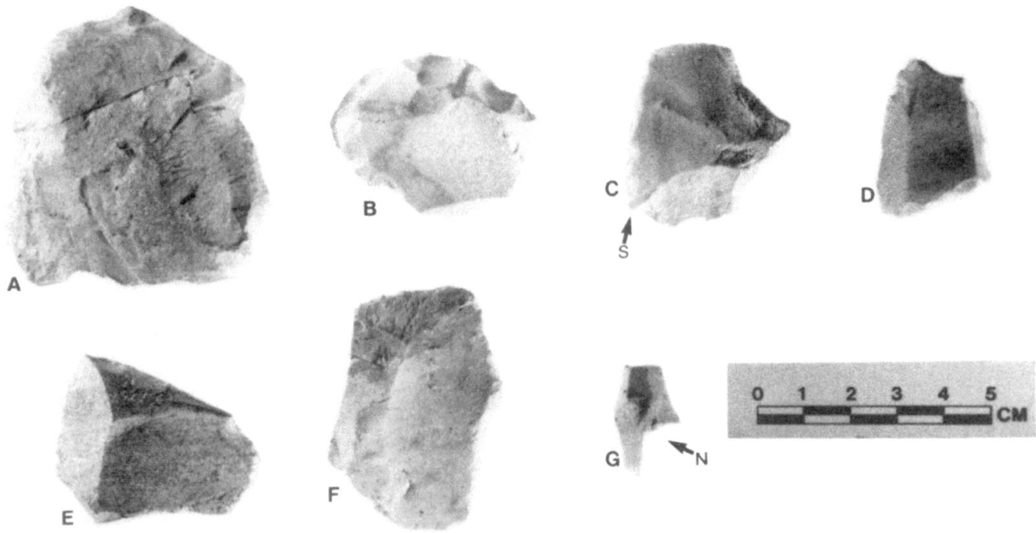

Figure 62. Other unifacial tools. **S** = spur, **N** = notch.

"hafted perforators" (Ellis and Deller 1988:119). Such items have been recovered from several other Parkhill complex sites. Overall, the distinct impression resulting from our analyses is that these Parkhill complex assemblages do not fit the widespread stereotype of a limited Paleo-Indian tool kit (see also Frison and Bradley 1980).

TABLE 38
Distribution of Blank Types by Tool Types

Blank Forms	Type or Class	N
Normal top corner	narrow end scraper	1
	Total =	1
Secondary top corner	proximal end and side scraper	1
	end scraper bit	1
	Total =	2
Normal side corner	offset end scrapers	3
	single concave side scraper	1
	single convex side scraper	1
	backed and snapped unifaces	3
	Total =	8
Secondary side corner	offset end scrapers	2
	combination notch/borer/denticulates	3
	backed and snapped unifaces	3
	other and fragmentary tools	2
	Total =	10
Top face	fluted points	N/A
	fluted end scraper	1
	proximal end and side scraper	1
	end and concave side scraper	2
	other end scraper	1
	end scraper bit	1
	single convex side scraper	1
	double convex side scraper	2
	piercer	1
	denticulate	1
	retouched flakes	2
	beaks	2
	Total =	15
Side face	triangular end scraper	2
	offset end scraper	1
	fluted end scraper	1
	narrow end scraper	1
	end and convex side scraper	1
	other end scraper	1
	single concave side with spurs	1
	single straight side scraper	1
	piercers	2
	denticulate	1
	retouched flake	1
	Total =	13
Flakes from biface cores	triangular end scrapers	7
	narrow end scrapers	2
	other end scrapers	1
	single concave side scraper	2
	single convex side scraper	2
	piercers	3
	combination notch/borer/denticulate	2
	retouched flake	1
	other scraper	1
	Total =	21
End biface thinning flakes	piercer	2
	retouched flakes	1
	Total =	3
Normal biface thinning flakes	piercer	1
	Total =	1
Channel flake	channel flake point	1
	Total =	1
Unknown blank forms	triangular end scraper	1
	narrow end scraper	1
	combination narrow/wide end scrapers	3
	proximal end and side scrapers	6
	other end scrapers	2
	end and convex side scraper	1
	single straight side scraper	1
	single concavo-convex side scraper	1
	double convex side scraper	2
	double alternate side scraper	1
	bend break tools	3
	combination notch/borer/denticulates	2
	piercers	9
	denticulate	1
	retouched flake	1
	transverse convex scraper	1
	notch	1
	"blocky" scraper	1
	Total =	38

Chapter 5

Flaking Debris

Some 1,010 pieces of Paleo-Indian flaking debris were recovered. This consists predominantly of Collingwood chert debris (94.5%) but it also includes a small amount of Bayport debris (5.5%) and one flake of the metasediment subgreywacke (see Kenyon 1979). It is possible that some of the Bayport flakes included in the sample are associated with later occupations of the site since at least two points diagnostic of these occupations (Fig. 4b, i) are on this material. However, most of this Bayport debris is probably Paleo-Indian associated. First, there are simply many more Paleo-Indian tools made of this material than are present in the later collections. Second, all of the Bayport debris conforms exactly in size and type to the definite and limited Paleo-Indian debris forms on Collingwood chert. Third, most of the Bayport debris and Bayport Paleo-Indian tools strongly cluster together in one area of the site. Moreover, in that area, there are no tools from Bayport chert representing the later prehistoric occupations. This will be demonstrated below in the discussions of spatial distributions.

The one subgreywacke flake is thought to be associated with the Paleo-Indian occupation because it was recovered from a probable Paleo-Indian feature. This raises the possibility that similar flakes in the collections are Paleo-Indian associated. However, since only four other flakes of this material were recovered, the exclusion of these other flakes should not seriously bias the analysis. As previous discussions have indicated, one Paleo-Indian Onondaga chert tool was recovered. Thus, it is possible that some of the small amount of Onondaga debris recovered is Paleo-Indian associated. However, because tools on this material are much more common among tools representative of later occupations and because the single Onondaga Paleo-Indian tool was recovered much outside the excavated area (in A-southeast; see Chapter 7), again little, if any, bias should be introduced by excluding this material.

The analysis of the debris generally follows the format developed for the Parkhill and McLeod site materials (Ellis 1979) but several refinements have been added through additional research and the benefit of hindsight. The analyses were designed to answer two basic questions. First, what types or categories of tools were altered on the site? This has a bearing on the nature of the activities at the site and is important in evaluating debris frequencies and debris-to-tool ratios as will be outlined below. Second, the debris analysis was directed towards determining what stages and steps in the alteration of lithic tools were carried out at the site. This is important in that the means by which the total reduction sequence is broken up into segments carried out at different loci is an excellent source of data on the organization of tool production systems and their articulation with other aspects of extinct cultural systems (see Collins 1975; Binford and Quimby 1972:346–47; Ellis 1984:16; Ericson 1984:3–7). Appropriately, the flaking debris was sorted into eight types and two residual categories thought to emphasize information that would answer these two questions.

The characteristics recorded for the debris and their means of measurement are essentially the same as those for the flake blanks listed in Appendix A. However, because of the very small size of most of the material, only somewhat subjective generalizations were possible for characteristics such as curvature, platform angle and banding orientation. Although all debris was sorted into the types and categories described below, in order to characterize each type, detailed data was collected only for a sample of the material (about 65% of the collection). The only exception was flake weight which was collected for all items.

Debris Types and Residual Categories

Information on the various flake types is given in Tables 39 to 51. Since detailed information is available in these tables, only generalizations will be presented in the following descriptions. Emphasis is placed on describing those attributes or variables which serve to distinguish either alone or in combination one type from another or which provide information relevant to the goals of the analysis given earlier.

Flakes Derived from Biface Reduction

Five flake types derived from biface alteration are recognized in the collections. These basically represent general stages of biface reduction, specialized operations such as fluting and in one case, apparent errors in flake detachment. The different stages of biface reduction with which certain flake types are equated are essentially those given in the discussions of point manufacture in Chapter 3: namely, thinning and finishing. There is no debris which can be equated with the stage of margin production.

Normal Biface Thinning Flakes

This type is represented by 28 examples (Fig. 63b). As the type name implies, these flakes were produced during the thinning stage of biface manufacture. They are quite large in relation to most of the flakes placed into other types but are much smaller than those flakes used as tool blanks which were referred to earlier as "flakes from biface cores." They always exhibit ground, faceted and acute-angled platforms, indicating removal from thin, highly prepared, bifacially chipped edges. The platforms usually exhibit a lip and rarely are isolated.

Lateral edges are consistently expanding. Curvature is usually symmetrical or distal and ranges from slight to pronounced. Those with more pronounced and symmetrical curvature suggest the bifaces altered were somewhat thick and narrow with at least one markedly convex face. Of the three flakes with no curvature, two have hinge terminations which were clearly unintentional on the part of the knappers.

The ventral surface is usually smooth. The dorsal surface exhibits parallel to convergent scars and indicates the removal of previous flakes of a similar size and configuration to the flake under examination. At the distal end of the dorsal surface, all complete specimens exhibit scars indicating considerable overlap with removals from the opposite biface edge.

TABLE 39
Collingwood Chert Flaking Debris

Debris Type	N	% of Total N	Total* Weight	% of Total Weight	Mean Weight
Normal biface thinning flakes	26	2.73	23.80	13.88	0.91
End biface thinning flakes	3	0.31	4.46	2.60	1.49
Biface thinning flakes in retouch	2	0.21	0.67	0.39	0.34
Biface reduction error flakes	5	0.52	1.12	0.65	0.22
Channel flakes	39	4.09	14.43	8.42	0.37
Biface finishing flakes	307	32.18	37.40	21.81	0.12
Normal uniface retouch flakes	146	15.30	21.24	12.39	0.15
Ventral uniface retouch flakes	7	0.73	1.44	0.84	0.2
Flat flakes	405	42.45	59.82	34.89	0.15
Other flakes	14	1.47	7.09	4.14	0.51
Totals	954	99.99	171.47	100.01	0.18

*Total weight of all items in type including fragments.

TABLE 40
Bayport Chert Flaking Debris

Debris Type	N	% of Total N	Total* Weight	% of Total Weight	Mean Weight
Normal biface thinning flakes	2	3.6	1.09	9.3	0.55
Channel flakes	2	3.6	1.19	10.2	0.60
Biface finishing flakes	12	21.8	1.77	15.1	0.15
Normal uniface retouch flakes	23	41.8	4.58	39.1	0.20
Flat flakes	15	27.3	2.83	24.2	0.19
Other flakes	1	1.8	0.26	2.2	—
Totals	55	99.9	11.72	100.1	0.21

*Total weight including fragmentary items in grams.

It is possible, as will be discussed later, that some of these flakes are a product of rejuvenation. However, the large size of the flakes suggests edge rejuvenation was not the sole goal of their removal. In particular, this size and their expanding nature indicates the removals were primarily intended to remove a large surface area of the biface and thus, to thin the item rather than to primarily refine plan outline shape. The considerable overlap with flakes removed from the opposite biface edge is also consistent with this inference as it suggests removals traveled far across the biface surface.

TABLE 41
Total Chert Flaking Debris

Debris Type	N	% of Total N	Total* Weight	% of Total Weight	Mean Weight
Normal biface thinning flakes	28	2.78	24.89	13.59	0.89
End biface thinning flakes	3	0.30	4.46	2.44	1.49
Biface thinning flakes in retouch	2	0.20	0.67	0.37	0.34
Biface reduction error flakes	5	0.50	1.12	0.61	0.22
Channel flakes	41	4.06	15.62	8.53	0.38
Biface finishing flakes	319	31.62	39.17	21.38	0.12
Normal uniface retouch flakes	169	16.75	25.82	14.10	0.15
Ventral uniface retouch flakes	7	0.69	1.44	0.79	0.21
Flat flakes	420	41.63	62.65	34.20	0.15
Other flakes**	15	1.49	7.35	4.01	0.49
Totals	1009	100.02	183.19	100.02	0.18

*Includes fragmentary items.
**Excludes one item of subgreywacke.

TABLE 42
Length, Width, Thickness and Weight of Flaking Debris

Length (mm)	N	R	X	S
Biface thinning				
normal	14	14–28	20.97	4.001
end	1	—	46.0	—
in retouch	2	15–16	15.5	—
Biface reduction errors	4	9–14.9	11.9	3.226
Channel flakes	—	—	—	—
Biface finishing flakes	82	4.2–13.0	11.7	3.340
Uniface retouch				
normal	101	4–16.5	8.6	2.849
ventral	5	4.5–13.0	8.8	3.947

Width (mm)	N	R	X	S
Biface thinning				
normal	11	16–29.0	20.2	4.244
end	3	14.6–21.0	17.2	3.347
in retouch	2	14–16.5	15.3	—
Biface reduction errors	4	9.8–11.0	10.3	0.526
Channel flakes	24	8–16.0	11.8	2.341

Table 42 (cont.)

	N	R	X	S
Biface finishing flakes	139	2.3–14.0	8.6	2.192
Uniface retouch				
normal	104	3–15.0	8.7	2.512
ventral	6	9.0–19.0	12.4	4.387

Thickness (mm)	N	R	X	S
Biface thinning				
normal	22	1.5–3.5	2.3	0.565
end	3	2–4	2.7	1.155
in retouch	2	—	1.5	—
Biface reduction errors	4	1–1.3	1.8	0.851
Channel flakes	24	1–2.5	1.8	0.330
Biface finishing flakes	171	0.2–2.4	1.1	0.388
Uniface retouch				
normal	105	0.4–5.0	1.5	0.811
ventral	6	0.8–3.1	1.6	0.944

Weight (gm)	N	R	X	S
Biface thinning				
normal	10	0.58–2.58	1.03	0.594
end	1	—	3.37	—
in retouch	2	0.27–0.40	0.34	—
Biface reduction errors	4	0.19–0.32	0.23	0.061
Channel flakes	—	—	—	—
Biface finishing flakes	77	0.01–0.39	0.14	0.094
Uniface retouch				
normal	101	0.01–1.13	0.15	0.474
ventral	5	0.04–0.63	0.24	0.251

*N = # of observations; R = range; X = mean; S = standard deviation.

Attempts were made to find sets of re-conjoined flakes removed from the same biface. However, no matches could be located.

Most of these flakes are probably from the thinning of bifaces intended to be reduced to fluted points since most of the bifaces from the site are fluted bifaces. In addition, on Collingwood chert items where the banding is visible, it almost always roughly parallels the longitudinal axis of the flake. This is to be expected on lateral preform thinning flakes as the vast majority of the points exhibit banding at right angles to the longitudinal axis.

TABLE 43
Platform Length and Width, Debris

Platform Length (mm)	N	R	X	S
Biface thinning				
normal	22	2.6–11.4	6.8	2.259
end	3	4.1–5.4	4.9	0.681
in retouch	2	4.5–5.9	5.2	—
Biface reduction errors	4	9.2–10.9	10.1	0.723
Channel flakes	6	3.8–4.4	3.9	0.316
Biface finishing flakes	165	1.1–6.6	3.9	1.229
Uniface retouch				
normal	101	0.7–11.8	2.7	1.501
ventral	6	3.5–10.0	6.5	2.632

Platform Width (mm)	N	R	X	S
Biface thinning				
normal	22	0.6–2.6	1.7	0.552
end	3	1.1–2.1	1.6	0.503
in retouch	2	1.2–1.3	1.3	—
Biface reduction errors	4	2.1–3.7	3.0	0.707
Channel flakes	6	0.8–1.8	1.3	0.372
Biface finishing flakes	176	0.3–2.4	1.1	0.394
Uniface retouch normal	101	0.2–4.0	0.7	0.474
ventral	6	0.4–2.2	1.5	0.618

*N = # of observations; R = range; X = mean; S = standard deviation.

TABLE 44
Platform Lip, Debris

	Present N	Present %	Absent N	Absent %
Biface thinning				
normal	16	72.3	6	27.7
end	—	—	3	100.0
in retouch	2	100.0	—	—
Biface reduction errors	4	100.0	—	—
Channel flakes	5	83.3	1	16.7
Biface finishing flakes	151	84.8	27	15.2
Uniface retouch				
normal	24	22.9	81	77.1
ventral	2	33.3	4	66.7

TABLE 45
Ventral Surface, Debris

	Smooth N	Smooth %	Undulations N	Undulations %
Biface thinning				
normal	18	81.1	4	18.2
end	1	100.0	—	—
in retouch	2	100.0	—	—
Biface reduction errors	4	100.0	—	—
Channel flakes	23	95.8	1	4.2
Biface finishing flakes	174	98.3	3	1.7
Uniface retouch				
normal	101	96.4	4	3.8
ventral	2	40.0	3	60.0

End Biface Thinning Flakes

Three end biface thinning flakes were recovered of which only one is complete (Fig. 63c). They exhibit acute, faceted, ground and markedly isolated platforms. The one complete flake expands rapidly from the platform to form roughly parallel edges.

Several characteristics serve to distinguish these flakes from those placed into other types. These include: a very large size; a lack of curvature; dorsal scars indicating the removal of previous large (in the normal biface thinning flake range) flakes transverse to the proximal-distal axis; and chert banding at right angles to the longitudinal flake axis.

The right-angled dorsal scars and banding and the lack of curvature leave little doubt that these flakes were removed from the end rather than the lateral edge of the biface. Again, given the right-angled banding and the predomination of fluted bifaces in the assemblage, we would suggest these flakes are from bifaces intended to be made into fluted points.

End thinning has been recognized for years in several industries such as the Achulean (Newcomer 1971:91) and has been reported more recently from several fluted point sites (Callahan 1979:15–19; Grimes 1979:114; Moeller 1980:71). End removals are often employed because they allow extensive thinning without reducing width. Also, they provide a means to remove abrupt hinge or step terminations from the biface surface (Callahan 1979:15–19; Muto 1971:59–63). Flakes detached from the same edge as a previous abruptly terminated

TABLE 46
Bulb Types, Debris

	Absent		Diffuse		Moderate		Pronounced	
	N	%	N	%	N	%	N	%
Biface thinning								
normal	10	45.5	8	36.4	4	18.2	—	—
end	3	100.0	—	—	—	—	—	—
in retouch	2	100.0	—	—	—	—	—	—
Biface reduction errors	2	50.0	1	25.0	1	25.0	—	—
Channel flakes	6	100.0	—	—	—	—	—	—
Biface finishing flakes	53	29.8	94	49.5	31	17.4	—	—
Uniface retouch								
normal	10	9.5	52	49.5	31	29.5	12	11.4
ventral	1	16.7	—	—	1	16.7	4	66.7

TABLE 47
Lateral Edge Configuration, Debris

	Parallel		Contracting		Expanding		Other	
	N	%	N	%	N	%	N	%
Biface thinning								
normal	3	13.6	—	—	18	81.8	1	4.6
end	—	—	—	—	1	100.0	—	—
in retouch	—	—	—	—	2	100.0	—	—
Biface reduction errors	—	—	4	100.0	—	—	—	—
Channel flakes	21	87.5	—	—	3	12.5	—	—
Biface finishing flakes	120	67.4	—	—	43	24.2	15	8.4
Uniface retouch								
normal	27	25.7	—	—	78	74.3	—	—
ventral	—	—	—	—	6	100.0	—	—

removal also tend to terminate at that step or hinge precluding adequate thinning. This can be corrected, however, by removing flakes from the opposite biface edge or an end. In the latter regard, it is notable that the one complete end biface thinning flake in the assemblage (Fig. 63c) has a large step termination on the dorsal surface near one edge produced by a previous lateral thinning flake removal.

Biface Thinning Flakes in Retouch

The two flakes in this type are almost identical to the normal biface thinning flakes. However, they are slightly smaller and exhibit a different dorsal scar pattern. Whereas the normal biface thinning flakes have large dorsal scars, the flakes in this type have small, roughly parallel-sided, scars. These flakes appear to have been detached after the application of a small, fine, parallel-collateral retouch similar to that found on fluted bifaces in the assemblage. This might indicate that the thinning flakes in retouch represent isolated removals in a subsequent finishing stage designed to correct errors in flaking. Alternatively, if preforms were retouched and used as is prior to manufacture into points (see below), these small dorsal scars may be indicative of the creation of use edges on the preforms; edges which would be subsequently obliterated by further thinning and reduction to points.

TABLE 48
Longitudinal Curvature, Debris

	Absent		Slight		Moderate		Pronouned	
	N	%	N	%	N	%	N	%
Biface thinning								
normal	3	18.8	4	25.0	6	37.5	3	18.8
end	1	100.0	—	—	—	—	—	—
in retouch	—	—	1	50.0	1	50.0	—	—
Biface reduction errors	2	50.0	—	—	2	50.0	—	—
Channel flakes	5	83.3	1	16.7	—	—	—	—
Biface finishing flakes	44	44.9	43 c	43.9	11	11.0	—	—
Uniface retouch								
normal	11	10.8	48	47.1	26	25.5	17	16.7
ventral	4	80.0	—	—	1	20.0	—	—

Channel Flakes

These flakes are a product of point fluting (Fig. 63a). After refitting, 41 of these flakes are represented. This total does not include three segments described earlier which fit onto channel flake segments adhering to preform tips removed by plunging flutes. Two of the channel flakes can be placed back onto fluting scars on the bifaces from which they were successfully removed.

All of the channel flakes are incomplete. Fourteen are platform ends while the remainder are distal or medial fragments. As a result of incompleteness, total length and weight cannot be determined. The sample examined in detail ranged from 8 to 35 mm in length and .09 to .91 g in weight.

Channel flakes exhibit several characteristics which serve to distinguish them from other types. Besides the fact they are long and thin, these include, first, a lenticular cross-section. Second, and as with the end biface thinning flakes, because they are removed from the base of the preform they almost always exhibit transverse chert banding and dorsal scars. An exception is one item from a double-fluted point where the scars are only partially transverse. Unlike the transverse scars on the end biface thinning flakes, those on the channel flakes are small and somewhat parallel-sided. A third characteristic of these flakes is a lack of curvature. Finally, as a product of the complex preparations necessary to ensure a successful flute removal, all of the platform ends are heavily ground, faceted and isolated.

TABLE 49
Curvature Placement, Debris

	Proximal		Symmetrical		Distal	
	N	%	N	%	N	%
Biface thinning						
normal	—	—	10	76.9	3	23.1
end	—	—	—	—	—	—
in retouch	—	—	2	100.0	—	—
Biface reduction errors	—	—	2	100.0	—	—
Channel flakes	1	100.0	—	—	—	—
Biface finishing flakes	1	1.9	28	51.9	25	46.3
Uniface retouch						
normal	—	—	26	18.6	65	71.4
ventral	—	—	1	100.0	—	—

Biface Finishing Flakes

This is the most common biface flake type. Of the 178 flakes of this type examined in detail, only 77 are complete. Indeed, completeness is one way of distinguishing these flakes from similar sized flakes such as uniface retouch flakes. The latter are almost always complete.

These flakes are very small. Platforms are usually ground and faceted, are invariably acute angled, tend to be biconvex in plan and are large in relation to overall flake size. Lateral edges are almost always parallel. Fur-

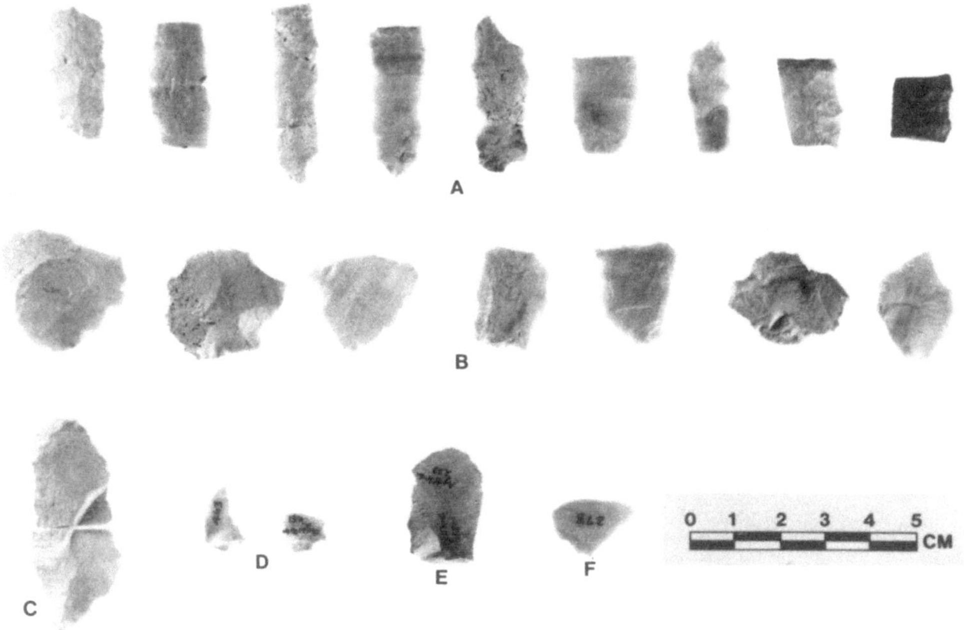

Figure 63. Flaking debris. **a**: channel flakes; **b**: normal biface thinning flakes; **c**: end biface thinning flake; **d**: biface reduction error flakes (ventral view); **e**: platform removal flake; **f**: ventral uniface retouch flake.

TABLE 50
Platform Surface, Debris

Flake Type	Faceted		Faceted & Scalar		Ground & Faceted		Ground Faceted & Isolated		Flat	
	N	%	N	%	N	%	N	%	N	%
Biface thinning										
normal	—	—	—	—	20	90.9	2	9.1	—	—
end	—	—	—	—	—	—	3	100.0	—	—
in retouch	—	—	—	—	2	100.0	—	—	—	—
Biface reduction errors	—	—	—	—	4	100.0	—	—	—	—
Channel flakes	—	—	—	—	—	—	6	100.0	—	—
Biface finishing flakes	14	7.9	—	—	164	92.1	—	—	—	—
Uniface retouch										
normal	3	2.9	—	—	—	—	—	—	102	97.1
ventral	1	16.7	5	83.3	—	—	—	—	—	—

thermore, of those with expanding lateral edges, many have little divergence (0–10°) and so, some 82.21% of the sample examined in detail have roughly parallel edges.

Ventral surfaces are smooth while dorsal surfaces exhibit one or two small parallel-sided scars. This suggests these flakes are a product of patterned, parallel, collateral flaking. In fact, although not attempted with the present assemblage, several overlapping pairs of these flakes were found in the Parkhill site collections (Ellis

TABLE 51
Termination Types, Debris

Flake Type	Feathered N	Feathered %	Hinge/Step N	Hinge/Step %	Broken N	Broken %
Biface thinning						
normal	12	54.6	2	9.1	8	36.4
end	1	33.3	—	—	2	66.7
in retouch	2	100.0	—	—	—	—
Biface reduction errors	4	100.0	—	—	—	—
Channel flakes	—	—	—	—	—	—
Biface finishing flakes	76	42.7	1	0.6	101	56.7
Uniface retouch						
normal	97	92.4	4	3.8	4	3.8
ventral	3	50.0	3	50.0	—	—

1979: Fig. 9) which left little doubt as to their production from this flaking pattern. A few of the complete specimens have dorsal-distal scars indicating slight overlap with removals from the opposite biface edge.

The small size of these flakes, the lack of significant overlap with removals from the opposite biface edge, and indications of patterned collateral flaking all suggest a concern with finishing the tool rather than with thinning. Since the points and preforms have a parallel collateral flaking which we would expect to produce flakes such as these, the flakes in the type are probably largely a product of point manufacture. However, some are undoubtedly from the retouching of biface tool forms other than points, and from resharpening.

Biface Reduction Flake Errors

This type includes only five flakes (Fig. 63d) which are small in overall size, especially length. However, they exhibit very large platforms with pronounced lips. In fact, most of the flake is platform. The platforms are always ground, faceted and acute angled. Another distinctive characteristic is that at least one, and usually both, of the lateral edges are contracting from the platform. Therefore, the flakes are widest at the platform edge.

It is suggested that these flakes are errors in removal. After the platform was prepared, the detaching "blow" did not initiate fracture close to the biface edge. Rather, it started a bend initiation deep on the biface surface. As a result, the edge collapsed and a short flake with a large platform was produced. This error could be produced in several ways such as striking the platform too far back from the biface edge, the presence of flaws in the raw material or the application of too much force to an inadequately prepared (i.e., too thin), platform. The overall frequency of this flake type suggests it was an uncommon error. Indeed, only one biface in the assemblage (Fig. 26d) exhibits an edge "bite" formed by this kind of error. The rarity of this flake type is not surprising given the acknowledged excellence of Paleo-Indian knappers.

Flakes Derived From Uniface Reduction

Normal Uniface Retouch Flakes

These distinctive flakes are derived during the "normal" alteration of uniface edges. They have been recognized in several industrial analyses (Frison 1968:150; Jelinek 1965; Shafer 1970:484; Witthoft 1952:479–80). There are 169 flakes of this type in the collection and most of these are complete. These flakes were derived by blows directed against the flat ventral surface of the tool blank. The flake removal traveled up the old or intended working face of the tool. Since most of the tools are unifacial, the platform is flat and approximates the flat ventral tool surface. The platforms tend to be right angled, are generally small, are often circular to irregular in outline and are never ground. Bulbs of force can be quite pronounced.

The dorsal surface exhibits roughly parallel scars. On this surface adjacent to the platform (i.e. at the old or intended working edge), a series of small, overlapping, hinged or stepped out flake scars are present (104 of 105). These can be interpreted to signify use of the tool prior to the flake removal although Brink (1978) has demonstrated that similar "wear" can result from manufacture.

Normal uniface retouch flake curvature tends to be more pronounced than that on biface finishing flakes and is generally distal in placement. Lateral edges are often expanding, largely because the flake removal can extend through the "back" of the tool (see below).

Based on flake termination, three varieties of these flakes were distinguished among the sample examined in detail. The first variety includes only four flakes. These hinged or stepped out in removal while the other varieties feathered out. The first of these latter varieties includes 69.1% of those flakes with feathered termination. In these cases, the removals traveled up the face and went beyond the extent of the previously beveled edge. They feathered out on the previously unretouched "back" of the tool. At the dorsal-distal end is a portion of this unretouched tool beyond the retouched surface. In longitudinal section, this facet appears at a right angle to the retouched surface. During flake detachment, the flake encounters the back segment, expands out along the natural ridge formed at the back segment juncture and abruptly curves.

The final variety includes those flakes with feathered termination which lack a back segment. These may be flakes which terminated short of the back or, alternatively, are from early in the tool's use-life before a well-defined beveled edge/back juncture was developed.

No concerted effort was made to refit these flakes onto unifaces in the assemblage. However, some matches of this nature were found at the Parkhill and McLeod sites (Ellis 1979: Figs. 19 and 20) confirming their removal from unifacial tool edges.

Finally, it should be noted that those items on Collingwood chert do not exhibit any patterning in banding orientation because the retouched uniface edge can occur at any angle with regard to this banding.

Ventral Uniface Retouch Flakes

Only seven of these distinctive flakes were recovered (Fig. 59f). They were detached by blows directed at the base of the old beveled edge of the uniface. This removed the flake body from the flat ventral surface (see Frison 1968:150; Shafer 1970:484). In effect, this turns the uniface into a bifacially worked object, albeit with only marginal retouch. The platform is faceted and exhibits a "scalar" surface of short, hinge or step terminated removals since it is part of the old working edge of the tool. As noted in previous discussions, this scalar effect may be produced through tool use. The dorsal surface approximates the old flat ventral surface of the tool and therefore, totally or partially lacks flake scars.

The platforms are right angled. Because the detaching blow was struck into a thick, flat surface, these flakes often have pronounced bulbs and undulations, are markedly expanding, tend to terminate in hinges or steps and lack longitudinal curvature.

The low frequency of these flakes suggests that inverse uniface removals were uncommon. This is confirmed by the fact that only one end scraper in the assemblage (Fig. 56h) exhibits an inverse retouch which would result from such flake detachments.

Residual Categories

Flat Flakes

The term flat flake is taken from DeVisscher et al. (1970:20). These flakes are the largest category numerically and by weight. They are simply thin flakes lacking platforms. Undoubtedly, these are the distal and medial ends of all of the other types with the exception of channel flakes which are readily identified even if they lack platforms. However, since the flakes from unifaces tend to be complete, most of these fragments are probably from flakes removed from bifaces.

Other

This category includes all other flaking debris. Including the large subgreywacke flake noted earlier, only 16 flakes fit into this category and this will deserve more discussion below. Although relatively complete, 14 of these flakes cannot be readily fit into the previously described types. The remaining two flakes can be described in more detail.

The first of these (Fig. 63e) is relatively large. This flake is notable in that it exhibits a bulb of force *on each face* at one flake end. These bulbs are associated with a separate platform on each face. It is quite clear that this is a flake removed during the removal of a platform and a bulb on a unifacial tool blank. In fact, one end scraper placed in the "fluted" type (Fig. 52h) exhibits a scar from such a removal designed to thin the proximal end for hafting purposes. Therefore, as a product of this form of removal, the dorsal surface of the flake described here exhibits the bulb and platform of the original tool blank's ventral surface while the ventral surface exhibits the platform and bulb resulting from the flake's own detachment. This flake is important because it indicates that some unifacial tools *were made on the site* or had more extensive reduction than simply the creation or resharpening of tool use edges. As previously implied, both the normal and ventral uniface retouch flakes are probably largely the result of rejuvenation.

The second flake is on subgreywacke. It weighs 10.1 g and exhibits a right-angled, heavily ground and faceted platform. This flake is important because it suggests nonsiliceous materials were sometimes (but rarely) employed by Paleo-Indian knappers in the Great Lakes region. This flake may have been used but the material type makes this difficult to determine with accuracy.

Discussion

Absence of Primary Manufacturing Stages

The whole sequence of lithic reduction was not carried out at the Thedford II site. For one thing, there are no cores *per se* in the assemblage. This generalization excludes the few point preforms whose reduction produced channel flakes and biface thinning flakes suitable for simple, expedient, briefly used tools. Second, the vast majority of the unmodified flakes are quite small. Ignoring the subgreywacke flake, only 15 flakes weigh over 1 g each and only one flake weighs more than 3.0 g. The few larger flakes are almost exclusively biface thinning flakes and several of these (e.g. Fig. 63c) had to be reconstructed from fragments as they collapsed during removal. Third, it is possible to assign almost all pieces of debris that retain platforms to types produced in the reduction of specific tool categories, such as biface or uniface. This indicates they are almost totally derived from the later stages of tool alteration, particularly retouching and resharpening.

The absence of evidence for primary manufacture is widespread on Paleo-Indian sites in the lower Great Lakes area. Almost all sites lack cores and yield only small debris. Examples include Parkhill (Ellis 1979), McLeod (Ellis 1979), Barnes (Voss 1977: Figs. 10–12), Holcombe (Fitting et al. 1966: Table 3), Ferguson (Ellis and Deller 1980: 103–4) and Gainey (Simons et al. 1984:268). Indeed, as both Deller (1979:7) and Wahla and DeVisscher (in Fitting et al. 1966:95, 102) imply, the small size of the waste flakes can be employed almost as a diagnostic indicator of Paleo-Indian sites in the area.

Sites in the area with evidence of primary manufacture are comparatively rare, but they include sites such as Fisher (Storck 1978:10) and Welling (Prufer and Wright 1970). Perhaps not surprisingly, these sites are strongly suspected to be in relatively close juxtaposition to lithic sources. This pattern of carrying out primary reduction only near lithic sources is very widespread in Paleo-Indian industries. It is common throughout the Northeast (see Funk 1982:xii; Gardner 1976, 1977) with source area sites such as West Athens Hill (Funk 1973), King's Road (Funk et al. 1969:11), Thunderbird (Gardner 1974, 1977), 6LF21 (Moeller 1980) and Williamson (McCary 1951) having considerable primary debris such as cores and large unmodified flakes. In contrast, many other sites which are all at some distance from the lithic sources employed (c. 30–320 km) yield almost solely small retouch flakes. These would include: Debert (MacDonald 1968:109), Twin Fields (Eisenberg 1978:136), Shoop (Witthoft 1952; Cox 1986) and Vail (Gramly 1982). Even some sites in quite close juxtaposition to lithic sources such as Shawnee-Minisink indicate that most primary manufacture was completed prior to the site occupation (Marshall 1981:65–73). This same pattern appears to occur elsewhere. For example, it is probably characteristic of much Folsom lithic reduction in the West (see Judge 1973:166; Tunnell 1975:9).

Debris-to-Tool Ratios

One might also employ the Thedford II debris-to-tool ratios (by frequency or weight) to argue for the absence of the primary manufacturing stages. Excluding materials outside the excavated area or with imprecise provenience, this ratio by frequency is 7.8:1 or alternatively 4.9:1 (if tool fragments are excluded or included respectively—different investigators have used each measure). However, these ratios may be slightly higher if some artifacts from a tool cache are excluded or if discrete Thedford II clusters are evaluated separately (see Chapter 7). The ratios at Thedford II are generally considered low ratios by investigators of Paleo-Indian sites (Eisenberg 1978:133, 136; MacDonald 1968:32, 34; Wilmsen 1970:76) but they do not reach the extremely low ratios reported for some sites such as Shoop (Witthoft 1952:471) or Vail (Gramly 1982:21).

However, we have several major reservations about the use of these ratios, especially in between-assemblage comparison. First, recovery techniques heavily influence these ratios. For instance, Hemmings (1970:103) has shown that despite an emphasis solely on tool maintenance at the Murray Springs Clovis site, an extreme emphasis on the recovery of fine items produced ratios on the order of 300–500 to 1. Therefore, it seems likely that the oft-commented low ratio at Shoop (estimated at 2–3:1) is partially a product of the fact the site was only surface collected (Ellis 1979:139) and that small waste was missed. Similarly, the low ratio at Vail in comparison to sites like Thedford II may be related simply to the more extensive use of finer (1/8″) mesh at the latter site.

A second factor influencing these ratios is the precise means by which these ratios are calculated. In some cases, they are only rough estimates (e.g. Wilmsen 1970). In other cases, and as noted above, tool fragments are sometimes included (MacDonald 1968; Gramly 1982) whereas in other cases, they are apparently excluded. Finally, in some cases debris from features is excluded (e.g. Gramly 1982) while in others such as at Thedford II and probably Debert, they are included.

A final major factor that must be controlled concerns the precise nature of the tools reduced on the site. For example, as Collins (1975:32) has noted, the reduction of one biface results in considerably more debris than the production of a single uniface. Moreover, as the analysis of the Thedford II material suggests, there is a greater tendency for the thin biface flakes to collapse in removal or to break by postdepositional disturbances such as plowing or screening as compared to the generally thicker uniface retouch flakes (see also Gramly and Lothrop 1984:134). Finally, it is notable that biface production on occupation sites is generally carried out to produce a tool form (fluted points) which is used off the site. As the number of complete points in kill sites attest, they are often discarded in that context rather than on the occupation site itself. Hence, on the occupation site, there may be lower numbers of bifaces discarded relative to unifacial tools which are often both made and discarded on occupation sites. Again, this would result in higher amounts of debris versus tools on sites where biface reduction was important.

The influence of this last factor is clearly evident at some sites. For example, this is undoubtedly a factor contributing to the high ratios at Murray Springs noted above where there is an emphasis on resharpening bifaces. Similarly, and more germane geographically, a good example is Grid or Area B at the Parkhill site. The debris from this site area is no different than that at Thedford II in terms of type or the size of the debris. Yet, the debris-to-tool ratio (excluding unidentifiable tool fragments) is estimated around 35:1. The obvious reason for this is that this site area is almost totally one of fluted biface manufacture, with 94% of the tools at this locus being bifaces (see Roosa and Deller 1982:12). The ratio of biface to uniface debris at Grid B is around 45:1 in comparison to the low ratio at Thedford II (2.3:1 in excavated area).

This Parkhill comparison allows the suggestion that whenever an assemblage is dominated by bifacial tools and preforms, we should expect a high debris-to-tool ratio. In fact, one can generally predict a low or high ratio for sites based on the low or high percentage of bifaces in an assemblage given relatively equivalent recovery techniques and sufficient sampling of an area by excavation. This may not produce exactly the same orderings of sites in terms of both biface percentages and ratios. Reasons for deviations in the orderings between the two will be discussed later. It means simply that the general ordering will be similar.

The Debert site (MacDonald 1968) provides a good example. Here, two areas (Sections D and E) have high debris-to-tool ratios (over 20:1) as compared to other areas which have ratios ranging from 1–7:1. Significantly, Sections D and E yielded the highest percentage of bifaces on the site (23.5% and 24.99%) while the other areas included only from 0.96% to 9.27% bifacial tools. As another illustration, the high percentage of bifaces at the Barnes site (around 70% of the assemblage; Voss 1977; Wright and Roosa 1966) would lead us to suspect a high debris to tool ratio and this is indeed the case (48:1). At the opposite end of the scale, we suspect that in addition to recovery techniques, this factor accounts somewhat for the very low ratio at the Vail site where bifacial tools (excluding *pièces esquillées*) account for less than 10% of the assemblage and where the ratio of biface to uniface debris is only around 2.1:1 (based on Gramly 1982: Table 1). A similar argument can be made for the Shoop site where bifaces make up only 12% of the artifact assemblage (based on Cox 1986).

The emphasis on biface versus uniface alteration will also affect the weight of debris versus tools. Since biface debris is produced in larger quantity, we should expect increases in the weight of debris versus tools where biface reduction was important. Moreover, since biface forms like biface thinning flakes and channel flakes are generally larger than uniface debris, an increase in biface alteration would result in greater weights of debris versus tools. Weights may however, be slightly more accurate than frequency counts since it allows some control over the fact that biface flakes tend to collapse in removal and obviates this factor's effects on the ratios. At Thedford II, the ratio of debris weight to tool weight is 1:4.7 for Collingwood chert, 1:17 for Bayport chert and 1:5.5 overall.

Initial Form of the Transported Raw Material

Given the absence of evidence for core preparation and tool blank production, we can be reasonably sure that lithic material was transported to the site in at least tool blank form. For unifacial tools, although some of these blanks were undoubtedly retouched or resharpened such that they appeared as side scrapers, end scrapers, etc., when brought to the site, there is some suggestion that these artifacts were often in the form of large unmodified flakes or tool blanks.

Admittedly, the flaking debris does not provide much evidence by itself for this practice. Only the one platform/bulb removal flake placed in the "other" category provides positive evidence. However, combined with the one graver or piercer made on the site by flake snapping and the two backed and snapped unifaces made on the same flake, it is probable that several of the intended unifacial tools were brought in as blanks. Indeed, the presence of a large number of unifacial tool blanks in the presumed Crowfield burial cache (Deller and Ellis 1984) and the presence of unmodified flakes in a fluted point associated cache at the Udora site (Storck 1981), suggest most of the intended unifacial tools were probably transported in blank form.

Some bifacial tools were probably in finished forms such as points or backed bifaces when brought to the site. However, the presence of channel flakes indicates that at least fluted bifaces were often transported as unfinished forms to locations of tool use. Based on two lines of evidence, it would appear that the transported forms were largely unfluted bifaces rather than totally unmodified flakes.

First, none of the biface thinning flakes from the site exhibit remnants of the original surfaces of tool blanks on their dorsal surfaces. This suggests the bifaces were somewhat thinned prior to transportation to the site. Second, and as at the Parkhill site, there are few unfluted preforms in the Thedford II assemblage and none from the earliest stage of reduction (i.e., margin production). If extensive reduction from flake blanks was being carried out, we would expect more "rejects" representative of these earlier stages.

One might also cite the low number of biface thinning flakes from the site as evidence of transportation of preforms rather than flake blanks even though, as will be discussed below, there are problems with this line of inference. In any case, there are 28 flakes in the assemblage from normal or end biface thinning. Flake scars on thinning stage preforms in assemblages from the area suggest the removal of at least 13 flakes of this nature per biface. Moreover, since these preforms only represent the final series of removals prior to discard, many more flakes would probably be produced per preform, the scars of which were obscured by later removals. Also, several more flakes could be detached had the preforms been reduced beyond the discarded state. If we assume: 1) that all of the flakes are from thinning preforms for fluted points; 2) that we have relatively complete recovery of these large flakes; and 3) that all preforms for points were transported as thinning stage preforms rather than a more reduced stage of reduction, then the number of thinning flakes appears very low compared to estimates of points made on the site.

For example, there are 14 Collingwood chert channel flake platform ends in the collection. Given two removals per point, then we would have to conclude that at least 7 of these points were produced on the site (as well as several others represented as preforms broken in manufacture upon which none of the channel flakes can be refit). Using only Collingwood chert thinning flakes, this would give a ratio of only 3.7 thinning flakes per point which is very low considering the number expected based on preform thinning flake scars. In short, using this line of reasoning, there are not enough thinning flakes in the assemblage to account for the complete production and thinning of these tools. Thus, transportation in more reduced forms than simple flake blanks is suggested.

The low frequency of biface thinning flakes might even suggest that the preforms were largely thinned prior to being carried to the site. In other words, the few thinning flakes represented might be simply isolated removals to prepare largely thinned bifaces for retouching and fluting. However, there are problems with this line of inference. In particular, it seems very possible that the thinning flakes represent sharpening of "preforms" which were used as biface tools prior to manufacture into points (see below). This would mean that individual preforms were not always extensively thinned at one sitting. Rather, they could be gradually reduced over time by the sharpening process. As a result, the number of thinning flakes produced in one sitting might be low.

In summary, while it seems evident that materials were transported as bifaces rather than as flake blanks, whether these bifaces were largely thinned or required more extensive thinning prior to retouching and resharpening is not clear. The latter alternative seems more likely, since the few thinning stage preforms in the assemblage would probably necessitate more thinning prior to fluting than simply isolated removals. Similarly, if the Crowfield site presumed grave cache data are applicable (see Deller and Ellis 1984), the latter interpretation seems more likely. The preforms (plano-convex bifaces) in the Crowfield assemblage would require more than simple isolated thinning flake removals to finish them into points.

Implications

Theoretical

The preceding discussions have several important theoretical implications with regard to interpreting debris assemblages from Paleo-Indian sites. First, some investigators have implied that debris size differences between sites (either alone or in combination with debris-to-tool ratios) can be used to infer precisely or relatively the time elapsed since lithic procurement episodes. The inference is that the more removed in time a site is from an episode of lithic procurement, the smaller will be the debris. In other words, a gradual distance-decay in flake size is assumed (Fig. 64a). However, as indicated above, this is probably a false assumption. There is probably an abrupt decrease in flake size as one moves away from lithic procurement epi-

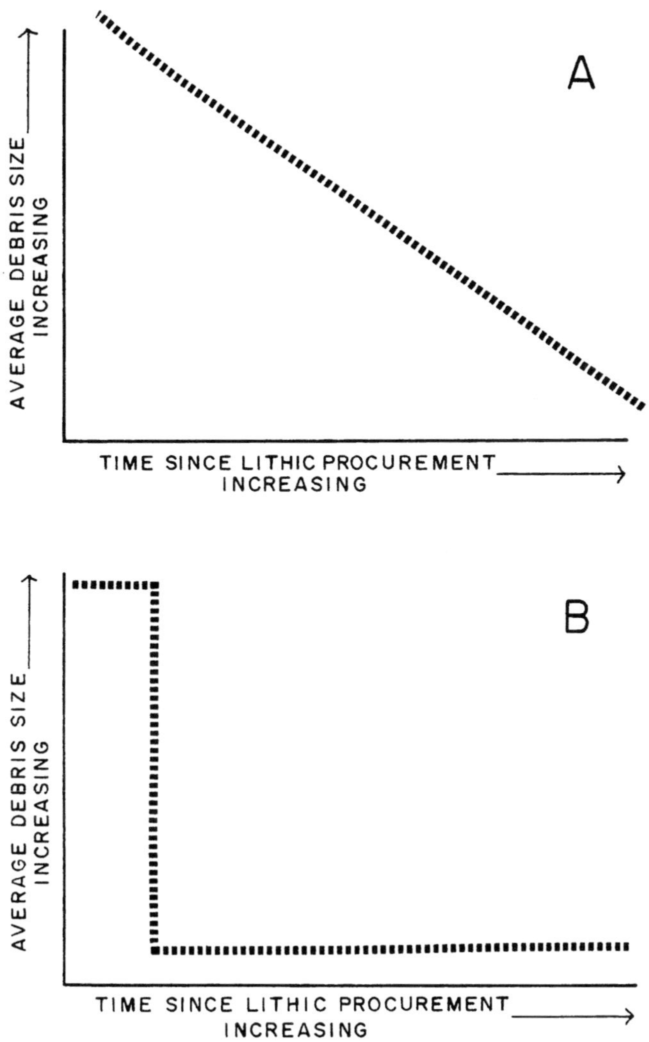

Figure 64. Alternative models of debris size characteristics with increasing distance to source.

sodes (Fig. 64b). As will be discussed below, this clearly represents a strategy on the part of Paleo-Indians to reduce the amount of lithic bulk transported away from sources which will be trimmed off and discarded as waste. Phrased another way, the small size of the debris is a product of *advance planning* on the part of Paleo-Indians rather than being a direct *situational product* of time elapsed since lithic procurement events.

Second, debris-to-tool ratios have also been used to infer time-elapsed since lithic procurement episodes. The inference is that the more removed in time a site is from these episodes, the smaller will be the ratio. This may be the case but unless one accounts for biasing effects such as variability in recovery techniques and whether bifaces or unifaces were being produced, these inferences must be approached with extreme caution.

Third, biface reduction obviously produces more debris. Since surface debris frequencies are often employed as a guide for areas to excavate, it would seem that on many Paleo-Indian non-quarry sites this would result in a bias towards areas of biface manufacture and use. As well, on sites where uniface alteration was important, low yields in test-excavations may lead to the mistaken conclusion that a site is not worthy of more extensive investigation.

Fourth, and again following from the idea that biface reduction produces more debris, density of debris in excavated areas has been used to isolate activity loci on some sites (see Voss 1977). Once again, simple density of debris will generally reveal only areas where biface reduction was important. In order to overcome this problem, it is necessary to classify biface or uniface debris in an assemblage and compare clusterings of these debris to overall debris densities. This plotting might reveal activity areas where, for instance, uniface reduction and use was important but not biface reduction.

Finally, we suggested above that recovery techniques being equal, one should generally expect increases in debris-to-tool ratios commensurate with an increase in the frequency of bifaces in an assemblage. However, this might not always be the case. Nonetheless, as will be shown in later sections dealing with internal spatial distributions at Thedford II, exceptions to this base-line model can provide important information on the nature of the occupations at various loci.

Substantive

The above discussions have indicated that the primary manufacturing stages of tool production, including core preparation and tool blank production, did not occur on the site. Rather, in addition to "finished" tools, lithic materials were transported to the site as flake blanks for unifaces and as bifacially worked preforms for tools such as points. It seems evident that this represents a strategy to reduce the lithic bulk transported to locations of tool use. As previous discussions have indicated, fluted point–associated groups were transporting lithic materials over large distances. By carrying out the most wasteful stages of manufacture at the quarry, only small debris of little utility such as uniface retouch flakes, would be produced away from quarries.

The most efficient means of reducing lithic bulk would be to transport only finished tools such as points, especially since fluting is often considered a wasteful process. In the case of points though, preforms were transported. There are several possible explanations for this apparent anomaly (see also Lothrop 1988, 1989).

First, the focus on a single source by Paleo-Indian groups in the area would suggest actual locations of tool use could be much removed in time and space from lithic procurement activities. Hence, precise predictions of actual tool needs away from quarries may have been difficult to make. The possibility exists that flexibility to meet various and relatively unpredictable tool needs away from quarries could be maintained in the tool kit by transporting unfluted biface preforms. Basically, the preform could be made into more than one functional tool type. For instance, Roosa (1977b:98–99) has suggested functional variability within the Barnes fluted point type. Certainly, this flexibility potential appears to be inherent in such preforms in other early industries such as Crowfield, where they could be made into at least three distinct tool types (Deller and Ellis 1984:51). Of course, this flexibility potential would be inherent also in tool blanks which could be made into several different unifacial tool types of probable use-significance.

Second, flexibility also could be achieved sequentially through the use-life of the tool. For example, it is possible, as suggested originally by Judge (1970, 1973) that the biface preforms could be used as is and later converted to fluted bifaces. In addition to introducing more flexibility into the system, this could also extend the use-life of a single piece of raw material. This "serial" use of a lithic item also would be of an advantage in transporting flake blanks for unifacial tools. For example, experimental (Walker 1978:713–14) and ethnographic (Hayden 1977:179) studies have shown that an unmodified (i.e., unretouched) sharp flake is more efficient in performing certain kinds of tasks than a retouched one. It is possible that the unmodified flake blank edge could be used for such tasks prior to being edge retouched into narrow end scrapers or other tools suitable for other tasks. It is just as possible, however, that the purposeful retouch on many tools is an actual product of resharpening rather a purposefully created working edge. Side scrapers are probably a good example here as implied earlier.

Third, the reduction of biface preforms and possible waste associated with their reduction could be minimized, as waste such as thinning flakes and channel flakes can be and often is employed for simple tools such as piercers, "flake knives" and channel flake points (again, see Judge 1973). This would avoid the use of larger blanks for simple tools—blanks which are often more suitable for the more complex tool types which require certain blank configurations.

Finally, even if the transported unfluted biface was not successfully thinned to serve as a point preform, it could be continually reduced for other tasks. Again, this would minimize waste. The "oval bifaces" in the assemblage are perhaps examples of this practice.

Summary

The analysis of the debris from Thedford II resulted in the recognition of several debris types significant in the stages of lithic manufacture and the general classes of tools reduced on the site. The debris assemblage is consistent with tool/preform assemblage from the site. For example, Bayport chert is poorly represented among both the tools and debris and there is little evidence of inverse uniface retouch among both classes of material. Similarly, there is little evidence of error in manufacture by biface edge collapse among either the debris or tools. Finally, the overall debris-to-tool ratio is low as is to be expected at sites where bifaces, and consequently biface reduction, are rare.

The analyses have indicated that the primary stages of lithic manufacture were not carried out at the site. Material was transported to the site in the form of finished tools, flake blanks for unifaces and as preforms for points. This is a general characteristic of all known sites in the Great Lakes area not known to be in the immediate vicinity of lithic sources. This system of tool production represents a strategy to reduce the bulk and maintain the flexibility of lithic items transported to locations of tool use.

Chapter 6

Features

Several prehistoric cultural features were located. The most definitive of these were 11 features associated with post–Paleo-Indian occupations of the site (Fig. 65). Most are probably of a Terminal Archaic age since diagnostic points of that affiliation predominate in the excavated area. Eight of these features (Features 1, 2, 4, 5, 6, 11, 12, 13A/B) are "structural"; that is, they had a clearly visible and well-defined outline in the soil. They consist primarily of generally oblong pits with shallow, basin-shaped profiles. These pits contained varying amounts of fire-cracked rock, some charred vegetal material (see Deller et al. 1986 for a summary), generally a few Kettle Point chert waste flakes and in one case (Feature 4), some deer and other unidentified large mammal bone. Occasionally, a diagnostic tool was included. For example, Figure 4e was recovered *in situ* in Feature 4.

The remaining three features (Features 7, 9, 14) are "artifactual." In sum, they consist of dense subsoil concentrations of lithic material which were not in any completely visible pit although, in two cases, partial outlines could be discerned. These artifactual features could be very deep (up to 55 cm into the subsoil) and appear to be specialized pits used in their final form as debris dumps. They included as many as 2,000 pieces of Kettle Point chert flaking debris. Occasionally, exhausted cores, hammerstones, expediently used flakes and biface fragments were included. We suggest that the absence of clearly visible outlines in the soil for these features may be partially related to an original paucity of organic materials.

In contrast to the above features, Paleo-Indian features were difficult to isolate and define. In fact, entirely definitive Paleo-Indian features were not located. Nonetheless, several probable features of this affiliation can be suggested and are described below.

The positive identification of these subsoil anomalies as Paleo-Indian features is made difficult by two factors. First, some of the probable features do not exhibit visible outlines in the subsoil. Rather, they are defined by the presence of subsoil artifact clusters. The presence or absence of feature outlines may be related to several factors such as the nature of the original contents, and the age and hence, extensive time available for outlines to be removed by leaching and other factors. Positive Paleo-Indian features lacking visible outlines are known to occur at other sites. Perhaps the best example is the large Paleo-Indian feature at the Crowfield site, the presence of which was indicated solely by a very dense subsoil concentration of heat-fractured artifacts (Deller and Ellis 1984). Furthermore, several Archaic features (such as the debris dumps noted above) are known to the authors which lacked discernible outlines in the soil. However, reliance on subsoil artifact clusters alone to delimit cultural features raises the possibility that these clusters were intruded in some instances by subsequent disturbances such as root or rodent activity, and this problem cannot be easily resolved.

A second confounding factor in identifying features—and this is especially a problem on sites with little or no subsequent matrix deposition—is simply that the sites have been subjected to thousands of years of root disturbances, rodent and microfaunal disturbances and subsequent cultural activities up to and including land clearing and farming by Europeans. These activities often hamper the delineation of features by intruding later material and blurring outlines. This is unfortunate but not unexpected. For example, at the Debert site (MacDonald 1968:19–20), which has more identified features than any other site of this age in the East, only the very bottom of relatively deep features were recognizable as cultural features and even these evidenced some disturbance by root/rodent activity.

The following describes the probable and possible Paleo-Indian features from Thedford II. The location of the most probable of these features is shown on Figure 72.

Probable Features

Feature 3

This feature was located at the juncture of 792N/790E and 792N/792E. It was linear and oblong in plan and was basin-shaped in profile (Fig. 66). It measured 250 cm long by 64 cm wide and a depth of 20 cm was preserved below the plow zone. The fill of the feature was a dark brown (10YR3/3) which contrasted with the surrounding yellowish-brown (10YR5/6) subsoil. The feature had been disturbed by root and rodent activity. During excavation, 5 Kettle Point flakes were found in these disturbances.

Paleo-Indian materials (excluding those from obvious disturbances in the area) included one tool fragment, one piercer or graver and 28 waste flakes. Fifteen of the waste flakes were recovered in the float samples. The waste flake collections included: 14 biface finishing flakes, 1 channel flake segment, a distal fragment of the only complete end biface thinning flake, and 12 flat flakes. Only one biface finishing flake of all the items recovered exhibits evidence of heat exposure.

Flotation samples were taken from the feature areas which did not appear to be disturbed. Charcoal from these floats included maple (*Acer* sp.) and beech (*Fagus* sp.). These were not expected given the inferred vegetation at the time of the site occupation.

The regular outline of the feature suggests it is of cultural origin. We attribute it to the Paleo-Indian occupation of the site because of the concentration of Paleo-Indian tools and debris in the area. This identification is predicated on the assumption that Paleo-Indian features will be associated with denser concentrations of artifacts as is the case on other known eastern sites with clearly defined features (e.g. MacDonald 1968; Gramly 1982). The feature coincides with a relatively dense concentration of flaking debris (Fig. 76) even if we include only materials recovered in the excavation techniques applied uniformly in all squares (screened plow zone, and flakes visible at troweled interface). In short, the concentration is not a product of the techniques used to excavate the feature.

Feature 8

This feature was located mainly in 800N/792E but one edge extended into 800N/794E. When first encountered, the feature was thought to be a tree throw. However, as the plow zone/subsoil interface was cleared and cleaned, several Collingwood chert flakes were uncovered (those present at interface are shown in Figure 67). These flakes were concentrated in the northwest portion of the feature. Two of the flakes (Fig. 67) were clearly outside the feature at interface but all subsequently recovered flakes were within the feature outline. This suggests, given their subsoil depth, that the two external flakes were plow-deposited or dragged. The flake concentration led us to treat the feature as cultural.

The feature had a slightly irregular outline in plan (Fig. 67) and measured 180+ cm long by a maximum of 134 cm wide. In profile views (Fig. 68), the feature was a rough basin, 42 cm deep, with three layers. The first, or uppermost, layer was restricted to a small area near interface and was a mottled silt. This layer was sterile and was a dark yellowish brown (10YR4/6) in color. It tended to intergrade with the third layer near the feature perimeter. The second layer was charcoal flecked and ranged from a very dark grey (10YR3/1) to a very dark greyish brown (10YR3/2). This ashy layer was highly disturbed from about the location of the B-B1 profile to the southeastern extremity. A complex area of root disturbances (including some small extant roots) and one rodent burrow were present in that area. These root disturbances resulted in a slight irregularity in the otherwise well-defined and regular A-A1 profile (Fig. 68).

The third layer, a red "halo," was a dark yellowish brown (10YR3/6). This was a definite layer rather than oxidized soil as it contained three Paleo-Indian tools and some flaking debris. This layer was mostly under the ashy layer although it appeared to be partially on top of the second layer near the northwest edge of the feature. This may be due simply to the difficulty in separating the halo from layer 1 with which it intergraded.

A final notable aspect of the feature was five small roughly circular stains visible at interface which disappeared completely about 5 cm into the feature. These may be postmolds. Their intrusion into the feature suggests they postdate it. They also raise the possibility that the feature is historic.

All cultural material from this feature was recovered from the second layer and the halo layer where it underlay the ashy layer at the feature edges. As previously implied, the southeast half of the feature was highly disturbed and so this was excavated and sampled for flotation separately from the northwestern half. This disturbed area produced 12 Kettle Point flakes and 6 pieces of fire-cracked rock which are probably intrusive. As well, charcoal from this area included beech and maple and one fragment of black walnut shell (*Juglans nigra*) which must be regarded as intrusive if this is a Paleo-Indian feature. The southeastern half also in-

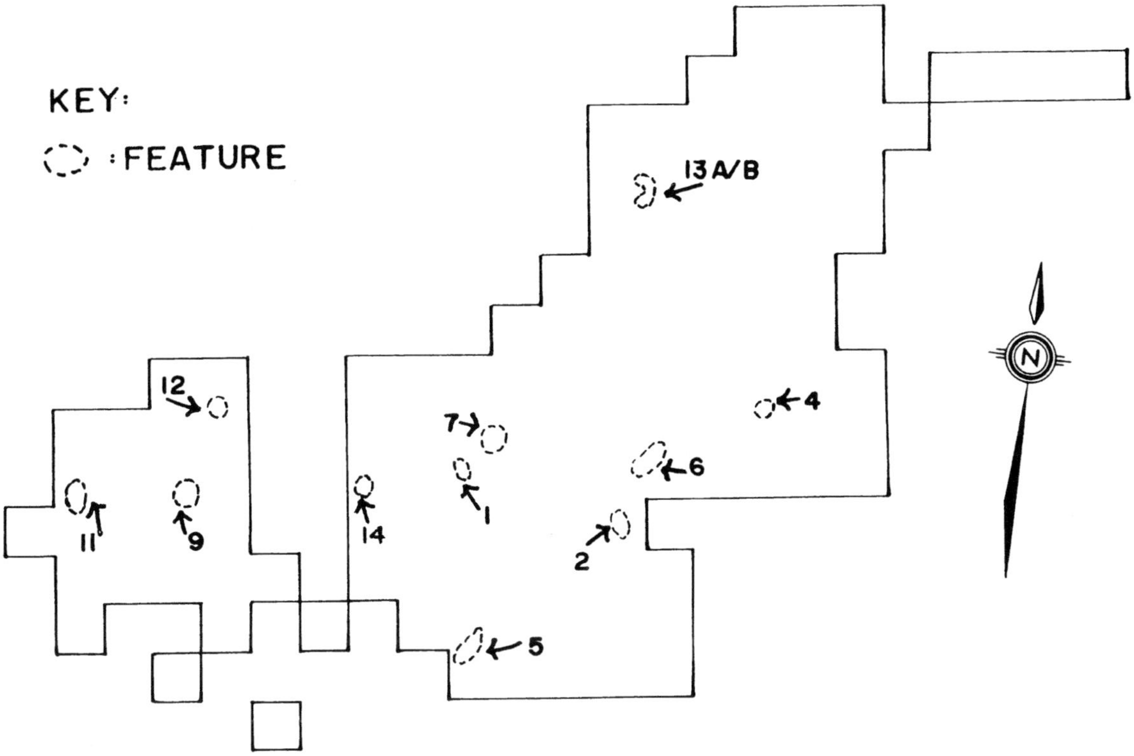

Figure 65. Distribution of non-Paleo-Indian prehistoric cultural features.

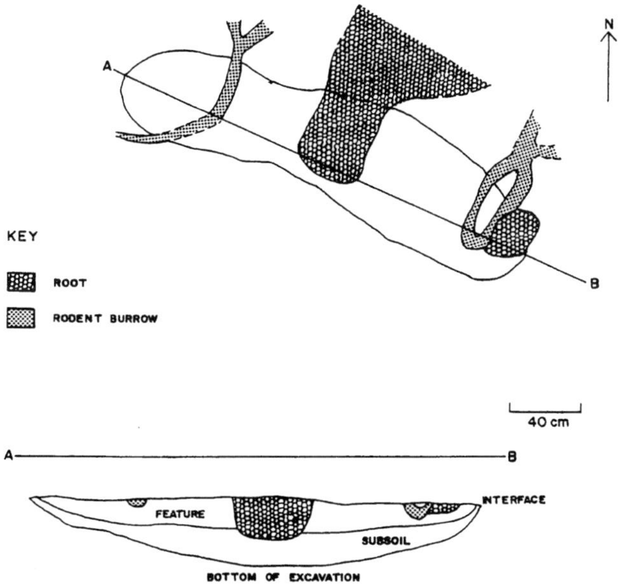

Figure 66. Feature 3, plan and profile views.

cluded 5 Collingwood chert tool fragments produced by heat exposure (one fragment is part of a single convex side scraper; the others are unidentifiable) and 8 Collingwood flakes (6 flat flakes, 1 uniface retouch flake and 1 unclassifiable flake).

By way of contrast, the northwestern half of the feature, which appeared largely undisturbed, lacked Kettle Point debris and fire-cracked rock. However, a single small piece (0.2 g) of charred maple and 2 very small pieces of uncharred root bark were recovered in the floats suggesting some disturbance. Some 27 Collingwood flakes, none of which were heated, were also recovered. These included 3 normal biface thinning flakes, 17 uniface retouch flakes, 6 flat flakes and 1 unclassifiable flake. It is significant that three Paleo-Indian tools were recovered from a small horizontal area at the southwest edge of the feature in the halo layer (Fig. 67). These three tools exhibited little vertical separation, occurring 13.5 to 19 cm below interface. All three tools were unheated. They included two multiple piercers or gravers (Fig. 60i, n) and one artifact placed into the

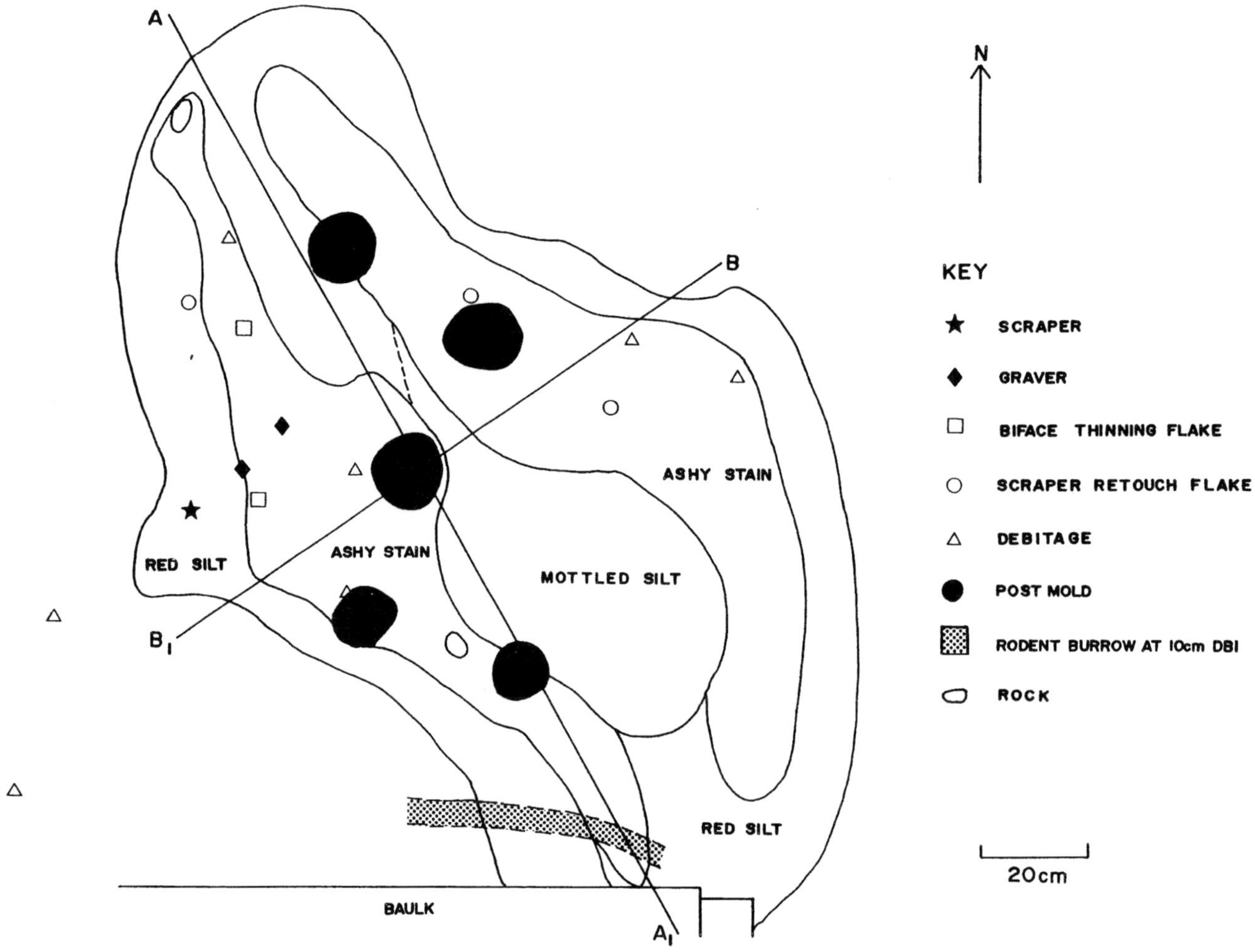

Figure 67. Feature 8, plan view.

"other" unifacial tool class. This last item (Fig. 62c) has a straight lateral and distal working edge and a piercer near the proximal end.

Interpretation of this feature is difficult. It has obviously been subjected to disturbance in the form of root/rodent activity. That the major evidence of this activity coincides with probable Archaic material (Kettle Point chert flakes) suggests the feature is originally of Paleo-Indian origin. Given the large amount of Kettle Point debris in the surrounding plow zone, the total absence of this material from the northwestern half of the feature is surprising. This does not rule out the possibility that the feature as a whole is a "tree throw" but it at least suggests the feature predates the Archaic activities.

A second line of evidence suggesting its Paleo-Indian affiliation is that there is a relatively dense concentration of flaking debris and tool fragments of Collingwood chert in the area (see Figs. 73, 75). Even excluding this material from the feature itself, the flaking debris recovered by techniques used in all squares still resulted in the recovery of materials in quantities matching the densest concentrations on the site as a whole. Finally, the three Paleo-Indian tools (all piercers) were recovered in relatively close juxtaposition in the largely undisturbed area. This close juxtaposition of three similar tool would appear to be due to more than simple chance.

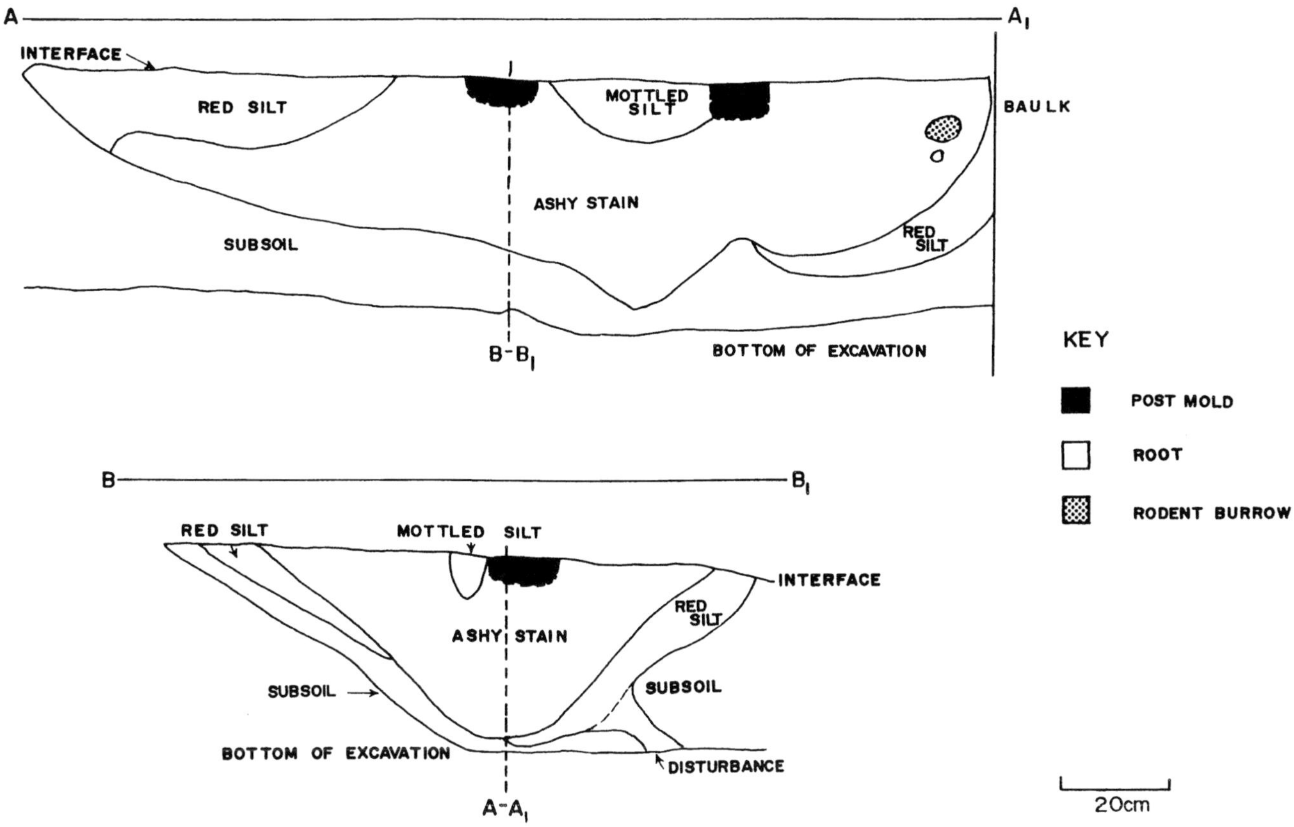

Figure 68. Feature 8, profile views.

Feature 10

This artifact feature was located at the juncture of squares 800 and 802N/802E. The feature is a subsoil concentration of Paleo-Indian tools and debris. There was no visible outline although presumably, a pit was once present. A large root stain was near the edge of the feature. The initial subsoil finds were assumed to have been intruded through root action even though these recoveries were clearly outside this feature. However, subsequent subsoil finds were well below the root suggesting a cultural feature. Materials were recovered to a depth of 29 cm below the plow zone and within a horizontal area of 1.5 by 1.5 meters.

The subsoil concentration included: a retouched flake (Fig. 61f), two single concave side scrapers (Fig. 57h, i), one of the two backed and snapped unifaces (Fig. 59a) which were made on the same flake (the other was in the plow zone just to the southeast), a tool fragment, a narrow end scraper (Fig. 53c) and 11 pieces of flaking debris. The debris included 6 biface finishing flakes, 3 flat flakes, a uniface retouch flake and the possibly used, subgreywacke flake noted earlier.

Feature 10 coincides with a relatively dense concentration of flaking debris (Fig. 76) but this is largely a product of the excavation techniques employed in the subsoil.

Feature 13A/B

These two interconnected features were located primarily in squares 808N/798E (Figs. 69–71). Feature 13A measured 60 cm long by 50 cm wide by 25 cm deep while Feature 13B was 48 cm by 26 cm by 20 cm. Two layers were present in the features and their location and characteristics are shown on Figures 70 and 71.

Initially, because of the presence of several root features in the area and a somewhat irregular outline, these connected features were thought to be a root disturbance. However, as the feature was taken down into

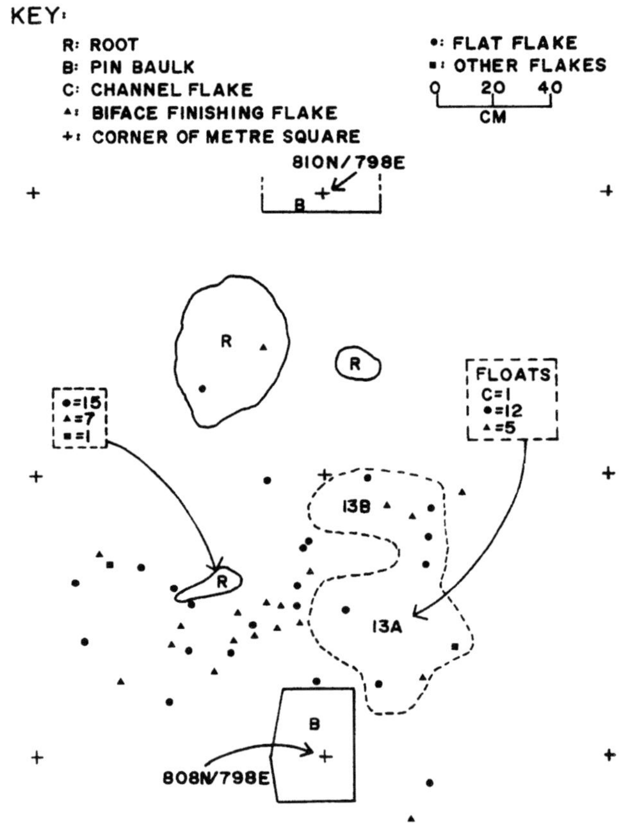

Figure 69. Feature 13A/B area.

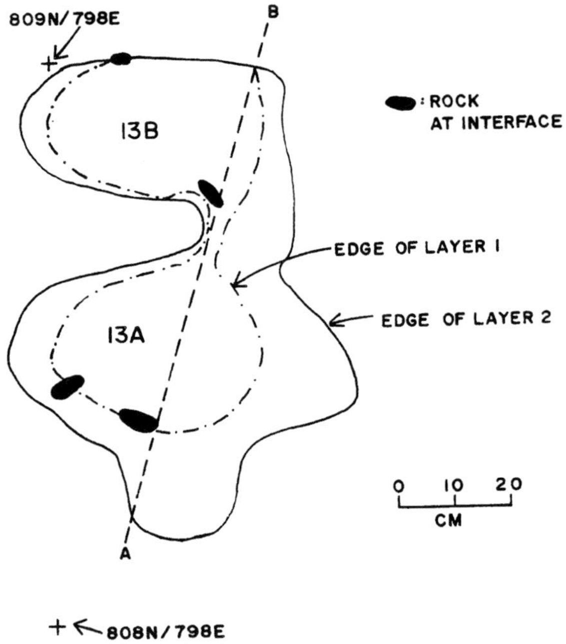

Figure 70. Feature 13A/B plan view.

the subsoil, the outline became more regular and resolved into the two small interconnected pits. It was clear that the somewhat irregular outline at interface was a product of deep plow scarring. The features definitely are cultural as it could be clearly seen where they had been dug down into the subsoil. Several Collingwood flakes and no Kettle Point debris were encountered within the features leading us to suspect a Paleo-Indian origin. However, there are several problems with accepting this cultural assignment.

First, several pieces of fire-cracked rock were found along the edges and this form of material has rarely if ever been reported for Paleo-Indian features. Second, the flotation samples included only beech (*Fagus grandifolia*) as identifiable charred material which would seem out of place in the inferred Paleo-Indian environmental context. It is possible however, that this material is intrusive given the number of root disturbances in the area. Third, the concentration of subsoil Collingwood chert debris was not restricted to within the feature. When plotted (Fig. 69), this debris clusters in a rough oval over an area of 150 cm by 80 cm and occurred up to 16.5 cm deep into the subsoil in areas outside the feature. While some of this material was clearly in root disturbances, most was not. Finally, a single notched point (Fig. 4c) was found near the base of the plow zone immediately above Feature 13A.

Given these anomalies, it is possible to interpret 13A/B in several ways. We suspect the very dense and obvious subsoil Collingwood chert debris cluster represents a former Paleo-Indian feature of a rough oval shape, the outlines of which have disappeared since its formation. Subsequently, the feature was disturbed by root action and probably, later prehistoric activities. Feature 13A/B itself may be a pit associated with those later occupations. A C-14 date on charcoal from the feature produced a date of 2120 ± 230 B.P. (I-13,025).

In any case, 86 pieces of Collingwood chert debris were recovered from the subsoil in and around Feature 13A/B excluding a Kettle Point flake and a Collingwood chert "popout" flake from a small root disturbance to the west. Twenty-eight of the flakes were found in the feature and the rest in the surrounding subsoil. The total sample includes 1 channel flake, 31 biface finishing flakes, 51 flat flakes and 2 unclassifiable flakes. In addition, a number of Collingwood flakes (53) were found in the plow zone above the subsoil concentration.

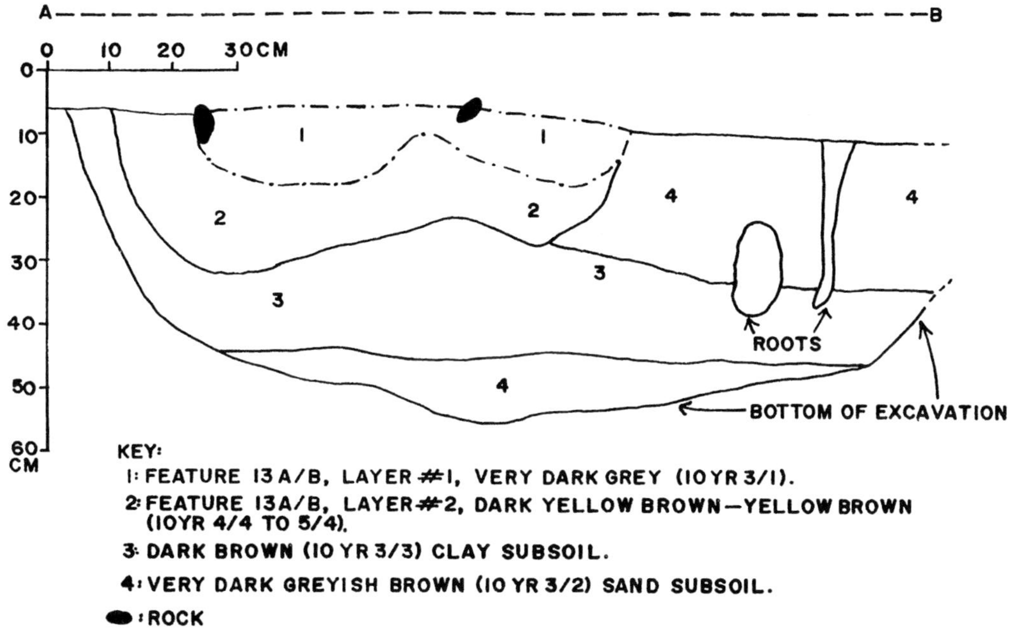

Figure 71. Feature 13A/B profile view.

Cache

A number of finds in the area designated A-northeast strongly suggest the presence of a Paleo-Indian tool cache. It had been subsequently disturbed by plowing and the digging of the tile trench system through this area (see Fig. 5). Most notable are several complete (except for recent breaks) fluted bifaces in essentially mint condition. While it is not uncommon to recover the occasional point or preform from fluted point sites which are complete and for which the reasons for discard are not clear (see Frison and Bradley 1980: 55–57), the number of these bifaces from this locus is truly exceptional.

The bifaces include four finished points (Figs. 20a-d; 21a-b), three of which are of Bayport chert. There are also four preforms (Figs. 22b-d, g, 23a-b) which could easily have been finished into points simply by grinding the lateral basal edges, by the application of a selective marginal bifacial retouch to regularize the edges and, in at least two cases, by providing a pointed tip.

These mint bifaces raise the possibility that other, more incomplete items from the area with recent breaks, are part of this cache. This would include: one point base (Fig. 20i), one preform base (Fig. 22f), one preform tip (Fig. 24b) and two fluted biface tips (Fig. 24c-d). The last two items are made of Bayport chert and, as noted in earlier discussions, a channel flake removed from one of these was recovered. Since the biface with matching channel flake may be part of the cache, the matchup cannot be used to infer that this is a preform.

The presence of a cache is also suggested by two other lines of evidence. First, all of the material strongly clusters in a relatively small area spatially, especially given site plowing. The plow zone material is centered around 810N/804E, directly above and around the tile drain (compare Figs. 5 and 72), and forms a notably dense cluster of artifacts when a density diagram of tools and fragments is constructed (Fig. 73). Second, and given previous discussions on the effects of biface reduction on flaking debris densities, the overall high percentage of bifaces in the A-northeast assemblage (especially preforms) would lead us to expect a relatively large amount of flaking debris and a high debris-to-tool ratio for the area in comparison to other site areas. However, as will be shown in the next chapter, this is not the case, probably because the tools were mostly not reduced at this locus. Instead, they were mainly brought to the site and cached as points and almost-finished preforms. This would greatly increase the number of bifaces at this locus but would not contribute much debris to the assemblage.

The presence of this cache might also suggest that some of the unifacial tools at A-northeast were cache contents. However, it is difficult to isolate possible examples as most unifacial tools do not differ from those in other areas. Perhaps the best candidate is a Bayport chert combination narrow/wide end scraper (Fig. 53f). This item is much larger than the other examples of this type from the site and therefore, may not be highly resharpened.

In summary, recoveries from A-northeast strongly suggest the former presence of a cache dominated by fluted bifaces. At least 8 and probably as many as 13 of these were cache contents. Presumably, the missing fragments created by recent breaks are in areas peripheral to those investigated. As well, they are probably in the fill of the tile trench. Since we wished to avoid interfering with the landowner's drainage system, only the top of the trench fill was removed and indeed, it did yield point fragments. Finally, we have knowledge of several large fluted biface segments being removed from the site by collectors.

Other Possible Features

At least two other possible features are suggested. Both are in A-west and lacked visible soil outlines. One was located at the juncture of squares 794N/776 and 778E and coincided with a dense clustering of artifacts and debris (compare Figs. 72, 73 and 76). This area was greatly disturbed by root action and the Paleo-Indian material was intermixed with later prehistoric materials. Paleo-Indian subsoil material, besides flaking debris and tool fragments, included an end and convex side scraper (Fig. 55c), a retouched flake (Fig. 61g) and a concave side scraper. Two fluted biface fragments were found in the plow zone above this subsoil concentration. As well, a narrow end scraper (Fig. 53d) was recovered right at the bottom of the plow zone in a plow furrow.

The other possible feature was in 792N/776E. It consisted of 18 Collingwood flakes (including two channel flake segments) and one Bayport flake but no tools. The Paleo-Indian material was not in any visible root or other disturbance even though some small roots cut through the area. These root disturbances contained 8 Kettle Point flakes. The subsoil Paleo-Indian debris was spread over an ovoid area measuring 50 cm by 80 cm and extended 30 cm into the subsoil.

Chapter 7

Spatial Distributions

Studies of microspatial distributions at Thedford II based solely upon the area excavated in 1981 have been carried out by Greco (1985). As well, Lothrop (1986) has undertaken a study of the distribution of more general artifact classes. Additional studies are planned for the future. Here, we will confine ourselves to a general overview and discussion of the distribution of artifacts and features at the site. Apparent patterning in the data is isolated and attempts are made to explain the contrasts and similarities between the various site areas.

The distribution of the tools and tool fragments within the excavated area, as well as the best candidates for Paleo-Indian features, are shown on Figure 72. This map shows the location of individual fragments, many of which can be cross-mended. It omits Bayport chert tool fragments which cannot be definitely assigned to the Paleo-Indian occupation of the site. Based upon this figure and a density diagram of tools and fragments (Fig. 73), it is clear that at least parts of four, major, distinct artifact concentrations were exposed in excavation. As noted earlier, they are referred to as A-west, A-centre, A-northeast and A-east. Plotting of cross-mended artifact fragments enhances the visual distinctiveness of these concentrations (Fig. 74) as the vast majority of the refits occur within the major clusters. The two exceptions are a single link between A-centre and A-northeast of fragments of a unifacial tool of unknown type or class and a single link between A-east and A-northeast of two very small basal fragments of a fluted point.

The inter-area links might be taken to indicate partial contemporaneity of the areas. However, in both cases, the refitted fragments are very small (under .5 g each) and were produced by heat exposure. While the initial heating and fragmentation of these tools may be due to Paleo-Indian activities, it seems unlikely that the subsequent transportation and separation of these fragments would be a product of such activities. Yet Gramly (1982), noting linkages of at least point fragments between discrete areas at the Vail site, has suggested these links indicate the sharing of meat between individuals resident at the discrete loci. This inference seems reasonable and could account for at least the fluted point link at Thedford II on the assumption the linked areas were produced by different social units (see below) and that game processing involved roasting which would fracture the tool. Furthermore, one could extend the argument to the unifacial tool link at Thedford II on the assumption this tool was used and broke initially in processing game. Nonetheless, we cannot rule out the possibility that plowing is the active agent of separation at Thedford II. In fact, these linkages between the major clusters extend north to south in relation to the grid orientation. This is the direction in which the site has been plowed and consequently, the direction we should expect most fragments to be separated. Similar patterns in relation to plowing direction have been clearly identified at other sites (e.g. Deller and Ellis 1984).

The four major clusters also can be distinguished on the basis of flaking debris density. Figure 75 suggests four major concentrations which generally correspond to the location of tool/preform clusters. There are some exceptions given more discussion below (for example, compare the distribution of tool and debris density in A-northeast; Figs. 73 and 74). For the purposes of these and other density diagrams to be presented later, cross-mended items from separate excavation units are counted once in each unit.

Figure 75 excludes Bayport debris (except for channel flakes) since there is evidence that later site occupants also used this chert. However, even if we include all Bayport debris because some of it may be Paleo-Indian associated (Fig. 76), a similar debris clustering is evident. These debris densities do suggest some sub-clus-

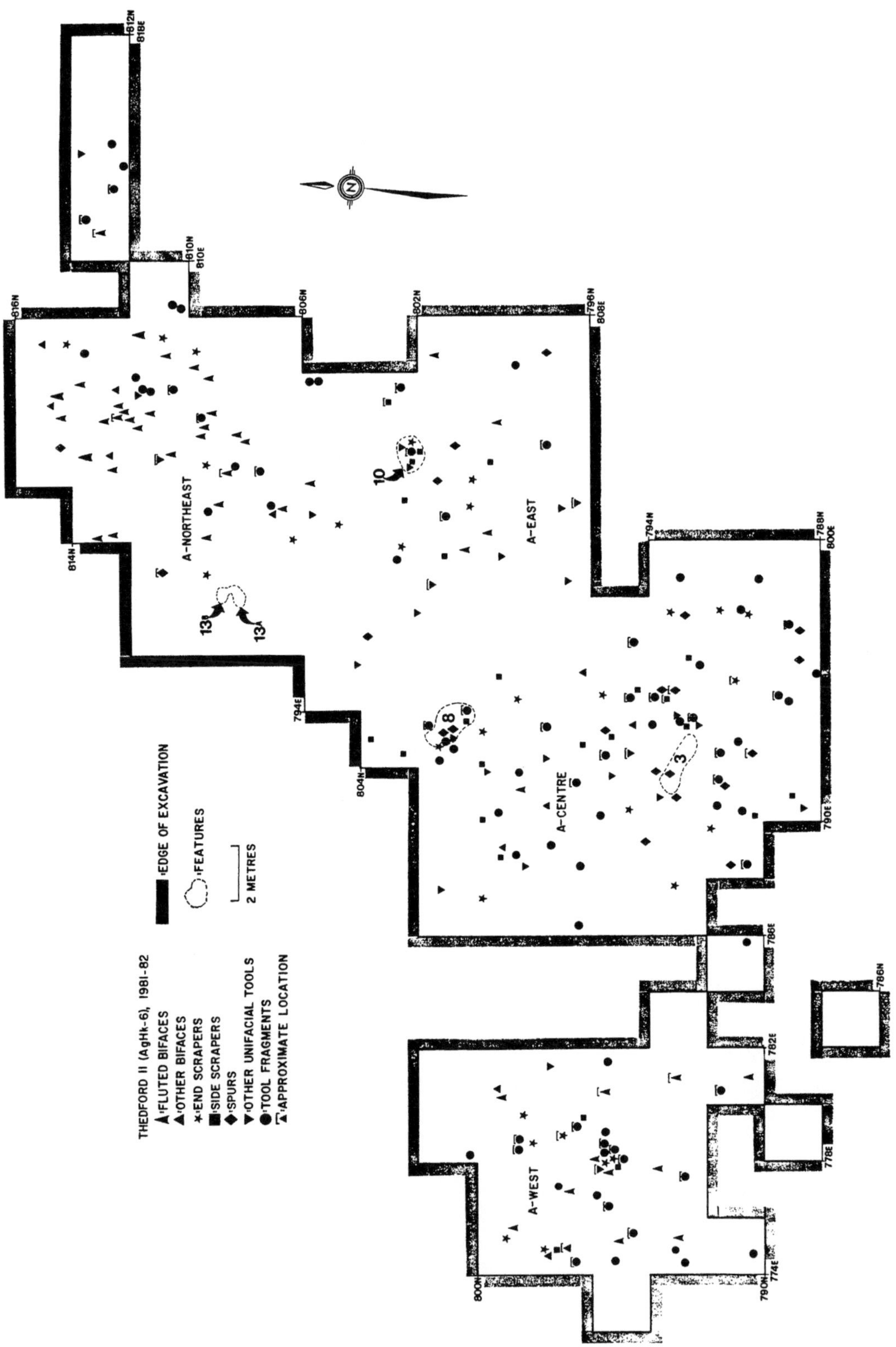

Fig. 72. Distribution of tools and tool fragments, Thedford II site.

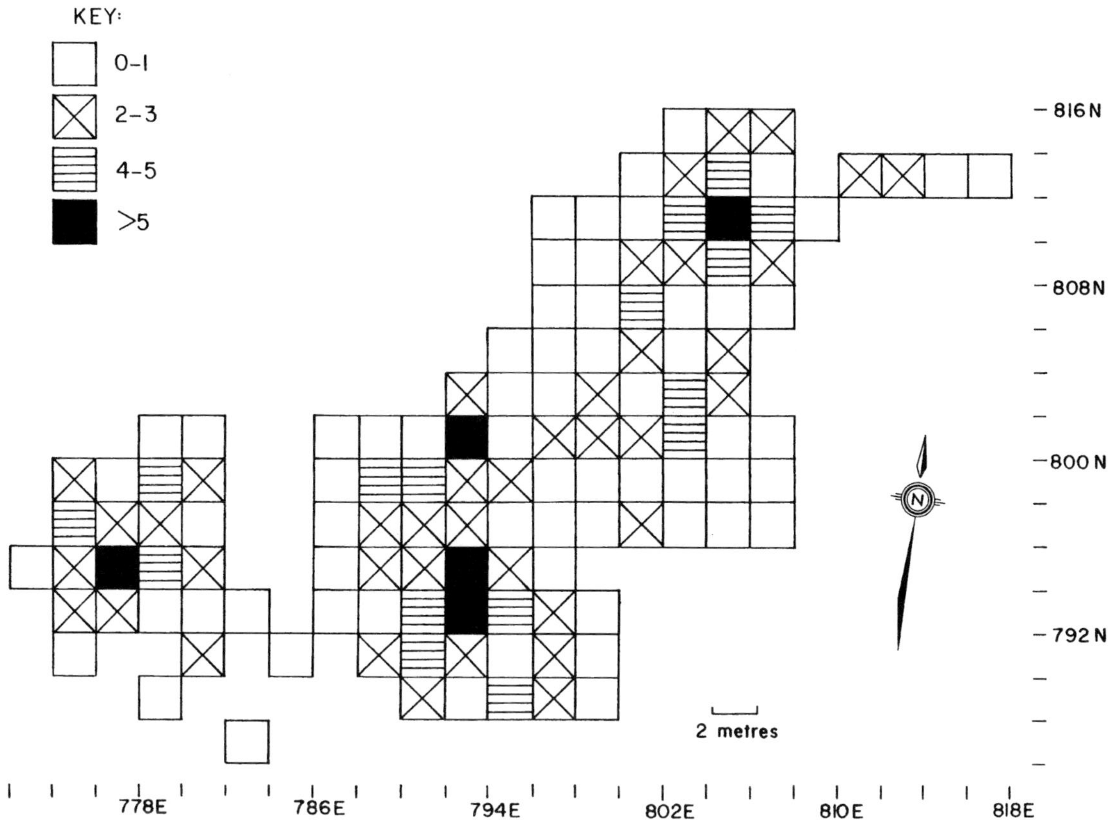

Figure 73. Density of tools and fragments.

tering within the major concentrations, in particular in A-northeast and A-centre. These will be discussed in more detail as each major cluster is described.

It is worth inserting that the clustering of all debris by classes and types, including Bayport chert items (Figs. 77–80), generally corresponds to overall debris densities. Uniface debris is especially notable in this regard (Fig. 80). In short, the clusters of all debris by site area are not simply a product of the locations of centers for biface reduction which, as noted in Chapter 5, we would expect to produce more debris.

In addition to the above clusters, surface finds suggest that two other clusters occur outside the excavated area. One of these is located just west of A-northeast. As stated earlier, that cluster was originally thought to be more distinct from the rest of Grid A based on the initial surface plottings in relation to bench marks and was referred to as Grid B. Yet, continued surface collection involving larger sample sizes and exact plottings by cross-tape triangulation to the datum points, as well as the more adequate delineation of other clusters through excavation, suggest Grid B is not that distinct.

The second unexcavated concentration (A-southeast) is located due south of A-east. Both this cluster and Grid B have not yielded enough material to draw conclusive interpretations. Hence, we will concentrate our discussions on the excavated clusters.

At the present time we conceive of the site as including two kinds of clusters. The first includes the excavated and unexcavated peripheral clusters (A-east, A-northeast, A-southeast, A-west and Grid B). The second consists solely of A-centre. In describing each excavated cluster below, the squares included in each for calculating debris frequencies per area are shown on Figure 81. Some squares indicated on this figure are omitted because they cannot be easily assigned in terms of waste flake content to one or the other concentrations of lithic material. These 13 squares contained only 22 flakes of which four are Bayport chert. The distribution of artifacts by each cluster are given on Tables 52 to 58.

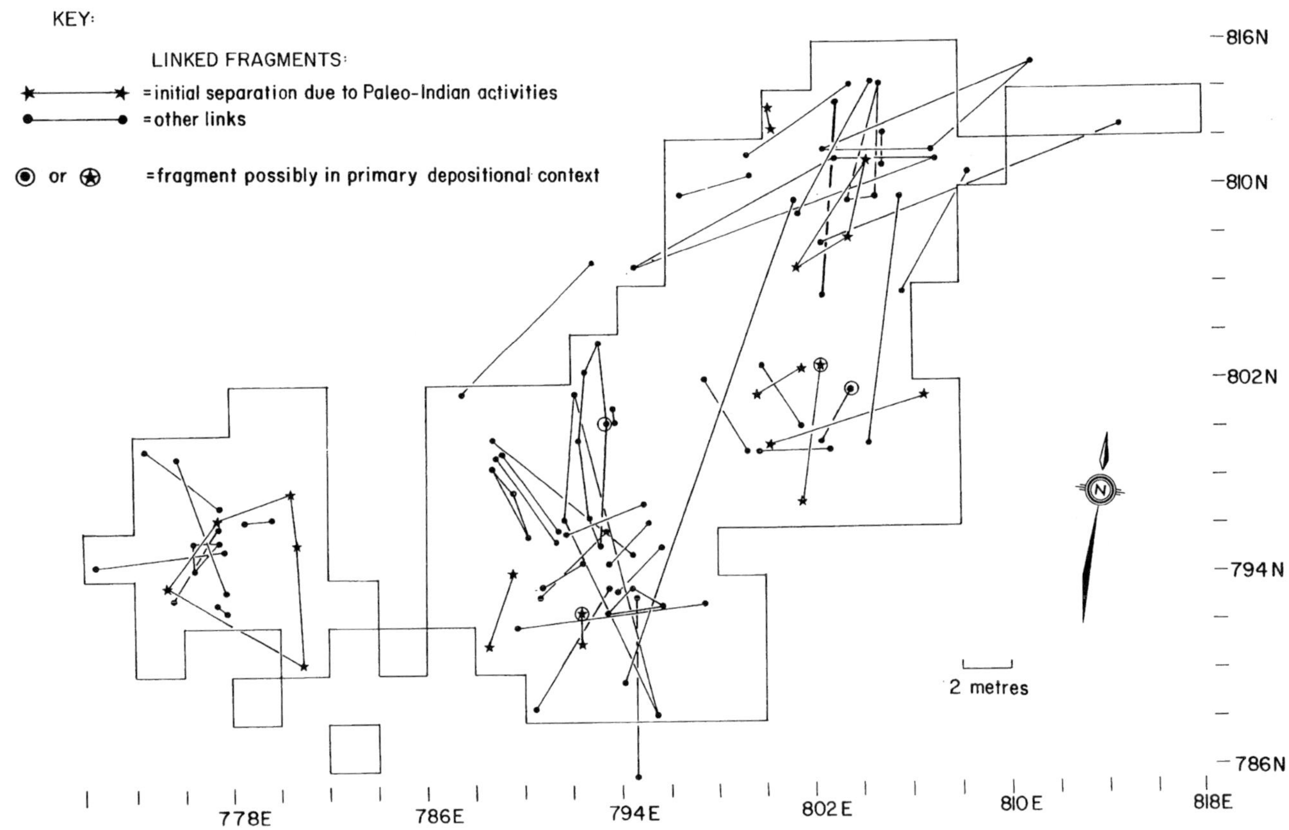

Figure 74. Cross-mended artifact fragments.

Peripheral Clusters

A-Northeast

The debris in this general area covers about 116 square meters. However, there is some question as to the integrity of the area as a whole (see below). As with most site concentrations, this must be regarded as an approximate estimate since the area probably extends into the unexcavated parts of the site. Yet, there is a tendency for almost all peripheral excavation units to have little in the way of flaking debris or tools/tool fragments based on density plottings (Figs. 73, 76). This suggests the size estimates for this and other clusters are fairly reliable.

Excluding plowing, postdepositional disturbances include heavy root action towards the western edge of A-northeast; later prehistoric occupations, as is especially evident in a concentration of Kettle Point chert artifacts and debris in the north edge of the area (Fig. 82; Concentration 2); and the tile drainage system which transects the area (Fig. 5). Perhaps significantly, patterns of cross-mends in the area (Fig. 74) show a tendency for fragments to be separated on a southwest-northeast axis at right angles to the tile drain. Therefore, we would suggest this tendency is largely a product of the tile drain's construction.

Thirty-seven tools or preforms can be associated with A-northeast. This includes two items, one of which is from the area but for which precise locational data is not available, and one of which is a surface find from just outside the excavated area. Since both items are or could be from outside the excavated area, these tools were not included in the calculated debris-to-tool ratios. It should be noted that a similar rationale was applied in producing these ratios for the other site areas.

A high percentage of the tools and preforms from the area (almost 40%; see Table 52) are made of Bayport chert, compared to the other site clusters (Tables 53–58; Fig. 83). There is a higher number of Bayport chert

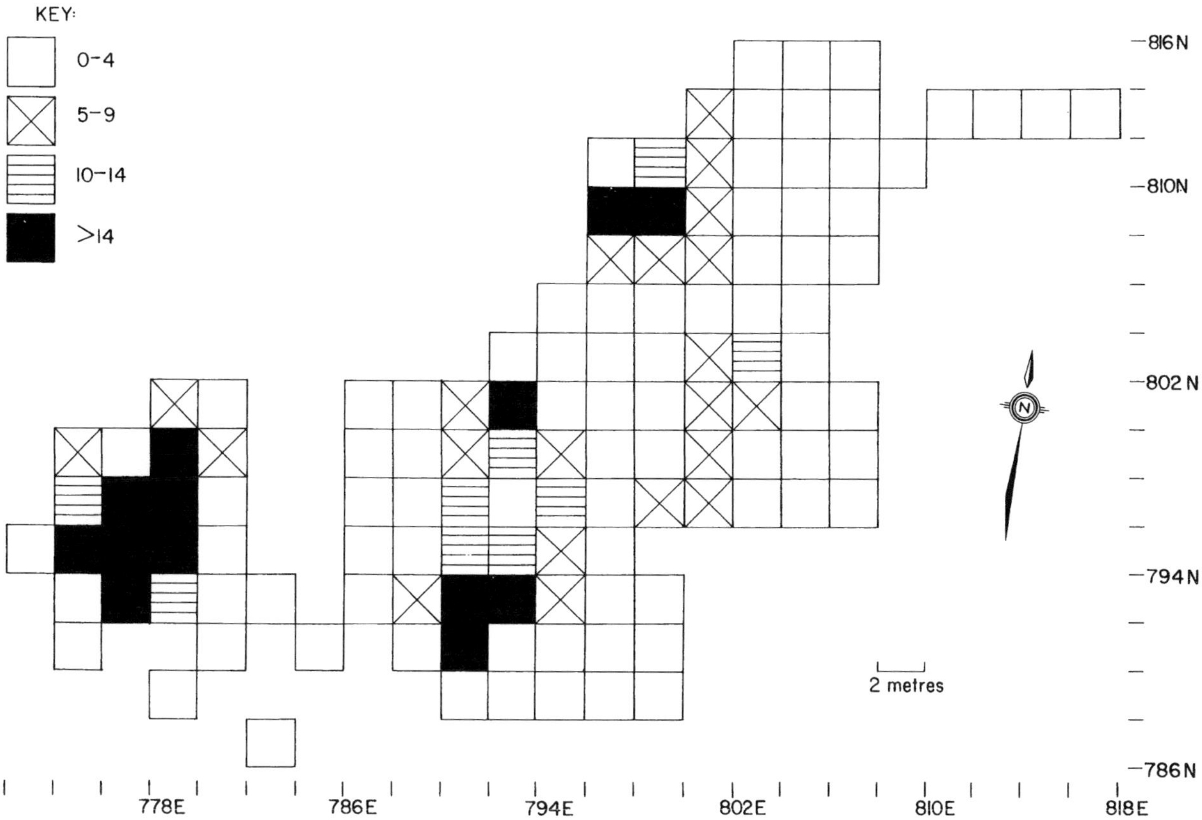

Figure 75. Density of Collingwood chert flaking debris.

flakes in this area (up to 6 flakes per square) in comparison to the other areas (although they still occur in relatively low frequencies; Fig. 84). This is consistent with the tool assemblage. Also, it strongly suggests that the Bayport debris from this area is associated with the Paleo-Indian occupation, particularly since no Bayport tools associated with the later occupations occur in this area.

The mechanisms by which the Bayport chert reached the Thedford II site are debatable. It has been suggested that the distribution of Collingwood chert as a major material on Parkhill complex sites is a product of a single "group" through time within which there is the maximal degree of consistent face-to-face interaction. Moreover, it has been suggested that sites dominated by Bayport chert in eastern Michigan represent a similar group in that area. Given these inferences, the consistent appearance of Bayport chert in small amounts on several Parkhill industry sites dominated by Collingwood chert must be due to either exchange or the movement of individuals between the groups represented by the two chert forms. At sites such as Parkhill (Roosa and Deller 1982), Dixon (Ellis 1984), Fisher (P. Storck: pers. comm.) and Mullin (Fig. 33, #5), the Bayport chert occurs in low frequencies compared to A-northeast (generally around 10%). This might suggest that movement of individuals is a more likely mechanism for the high percentage of Bayport chert at A-northeast.

Over half of the tools from A-northeast are fluted bifaces (Table 52; Fig. 85). While many of these are points and preforms in mint condition connected with a cache in the area, the presence of channel flakes, at least one preform broken in fluting and one fluted biface upon which a channel flake can be replaced (Fig. 24c), suggests that points were made in the area. In addition, there are two finished Bayport chert points broken by transverse snaps in projectile use as well as a Collingwood and a Bayport chert point ear. These basal seg-

Table 52
Artifact Total, A-Northeast

Type	Site Total	C	B	O	Area Total	% of Area Total	% of Site Total
Fluted points	12	3	6	—	9	24.3	75.0
Fluted preforms	12	6	2	—	8	21.6	66.7
Fluted bifaces	8	—	2	—	2	5.4	25.0
Unfluted preforms	3	2	—	—	2	5.4	66.7
Alternately beveled bifaces	5	—	1	—	1	2.7	20.0
Concave-base bifaces	2	1	—	—	1	2.7	50.0
Oval bifaces	2	1	—	—	1	2.7	50.0
Channel flake point	1	1	—	—	1	2.7	100.0
Pièce esquillée(?)	1	—	—	—	1	—	—
Backed biface	1	—	—	—	—	—	—
Triangular end scrapers	10	1	2	—	3	8.1	30.0
Offset end scrapers	6	—	—	—	—	—	—
Fluted end scrapers	2	—	—	—	—	—	—
Narrow end scrapers	5	1	—	—	1	2.7	20.0
Combination narrow/wide end scrapers	3	—	1	—	1	2.7	33.3
Proximal end and side scrapers	8	1	—	—	1	2.7	12.5
End and concave side	2	—	—	—	—	—	—
End and convex side	2	—	—	—	—	—	—
Other end scrapers	9	1	—	—	1	2.7	11.1
Side scrapers	15	—	—	—	—	—	—
Combination notch/borer/ denticulates	7	2	—	—	2	5.4	28.6
Bend-break tools	3	1	—	—	1	2.7	33.3
Backed and snapped unifaces	6	—	—	—	—	—	—
Piercers	18	1	—	—	1	2.7	5.6
Denticulates	3	—	—	—	—	—	—
Retouched flakes	6	—	—	—	—	—	—
Other unifaces	7	1	—	—	1	2.7	14.3
TOTALS	159	23	14	0	37	99.9	23.3

C = Collingwood; B = Bayport; O = Other.

ments lacking tips suggest broken point fragments were removed from the hafts and perhaps, new points were rehafted in their place.

The remainder of the tool types and classes largely occur in low frequencies as they do throughout the site. However, it is notable that 25% of those tool forms at the site (28.6% in excavated areas) incorporating narrow end scraper bits (either alone or in combination with other working edges) are associated with A-northeast. It is also of note that side scrapers are totally absent and that only one piercer or spur (in two fragments) is from the area despite the fact piercers are the most common unifacial tool type at the site (Table 52; Fig. 87).

Flaking debris is present in moderate amounts, averaging about 9 flakes per two-meter unit (Table 59). The ratio of biface to uniface debris is the highest per site area and this is in line with the ratio of bifacial to unifacial tools which is also the highest per site area (Table 60). However, the debitage-to-tool ratios are relatively low (Table 60) compared to the other site areas, particularly A-west. In other words, given that this site area has the highest percentage of bifaces, it should have the highest debris-to-tool ratios based on previous arguments. The obvious explanation for this inconsistency is the inference that the number of bifaces from the area is augmented by fluted bifaces associated with the cache. These bifaces were discarded but largely not altered at this locus.

It should be emphasized that the debris-to-tool ratio and biface-to-uniface debris ratios from this site area are

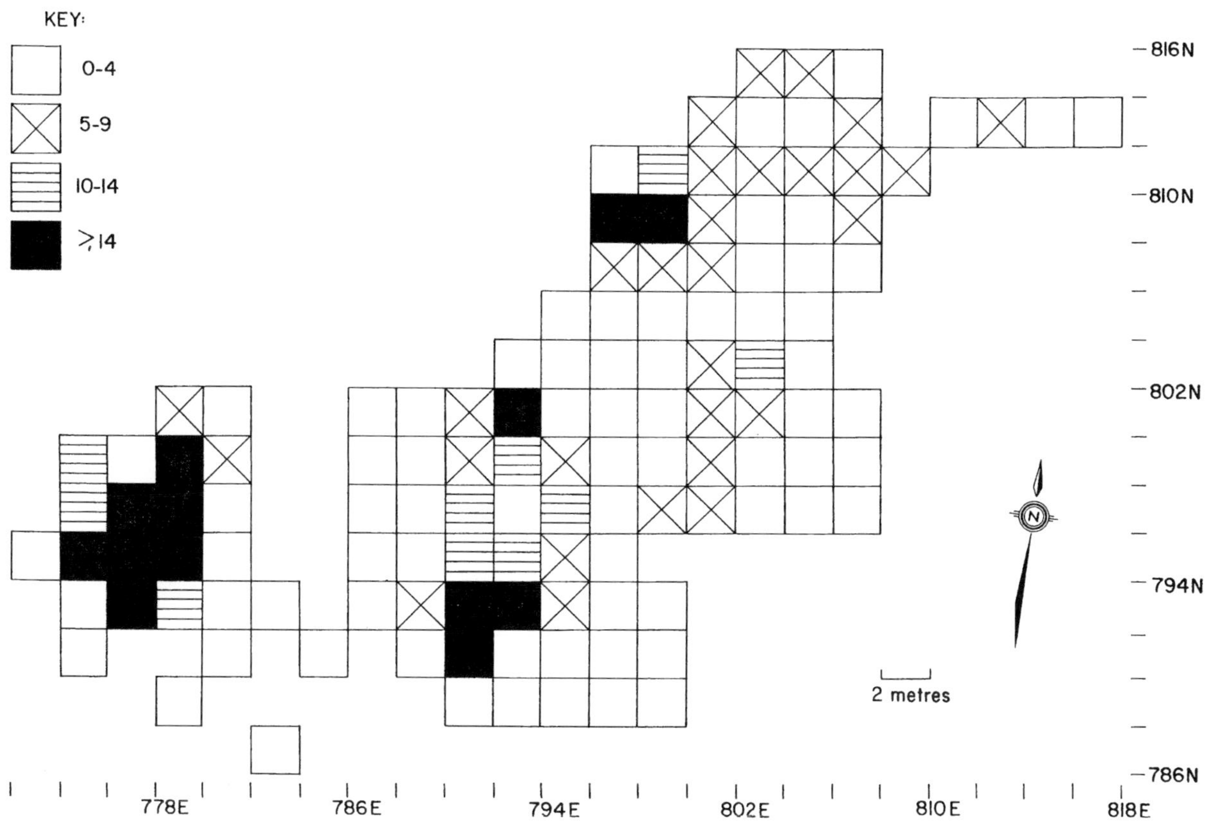

Figure 76. Density of all Paleo-Indian flaking debris.

probably much lower than the totals given on Table 60 since much of the debris is probably not associated with the tool assemblage. In particular, there is some question as to whether A-northeast can be treated as an integral unit. Two sub-clusters are apparent in A-northeast: one generally to the west of the 800E line (shown as "west end" on Fig. 81) and one to the east of this boundary. The eastern portion contains most, if not all, of the tools, tool fragments and preforms from A-northeast (Fig. 72). Conversely, it is very low in terms of waste flakes. In fact, only 40% of the waste flakes occur in the eastern segment, despite its much greater spatial extent. Also, the waste flakes from this eastern area include all those of Bayport chert as well as, of course, all the Bayport chert tools. In addition, the waste flakes in the eastern area generally did not originate in biface manufacture, as the ratio of biface to uniface debris (using the 800E line as an arbitrary dividing point) is a low 1.4:1. This markedly low ratio, when compared to the number of mint fluted bifaces from the area which are co-extensive with this debris, again strengthens the idea of a cache in this area. However, some point manufacture did occur in the eastern area as it includes seven channel flakes, a point-channel flake cross-mend and preforms broken in manufacture.

In contrast to the eastern half of A-northeast, the western half has few or no tools/preforms and has a dense concentration of debris associated with Feature 13. This includes considerable evidence of biface manufacture such as several channel flakes. None of these channel flakes refit with points to the east. This, along with the total absence of Bayport flaking debris, suggests that the west half and, in particular the Feature 13 flake cluster, is not associated with the rest of A-northeast.

We suggest that Feature 13 is associated with Grid B, tools from which occurred as surface finds within four meters to the northwest of the feature. Alternatively, the Feature 13 cluster may be "extraneous" and not clearly associated with any defined major tool con-

TABLE 53
Artifact Total, A-East

Type	Site Total	C	B	O	Area Total	% of Area Total	% of Site Total
Fluted points	12	—	—	—	—	—	—
Fluted preform	12	1	—	—	1	4.8	8.3
Fluted bifaces	8	1	—	—	1	4.8	12.5
Unfluted preforms	3	—	—	—	—	—	—
Alternately beveled bifaces	5	—	—	—	—	—	—
Concave-base bifaces	2	—	—	—	—	—	—
Oval bifaces	2	—	—	—	—	—	—
Channel flake point	1	—	—	—	—	—	—
Pièce esquillée (?)	1	1	—	—	1	4.8	100.0
Backed biface	1	—	—	—	—	—	—
Triangular end scrapers	10	—	—	—	—	—	—
Offset end scrapers	6	1	—	—	1	4.8	16.7
Fluted end scrapers	2	1	—	—	1	4.8	50.0
Narrow end scrapers	5	1	—	—	1	4.8	20.0
Combination narrow/wide end scrapers	3	—	—	—	—	—	—
Proximal end and side scrapers	8	—	—	—	—	—	—
End and concave side	2	—	—	—	—	—	—
End and convex side	2	—	—	—	—	—	—
Other end scrapers	9	—	—	—	—	—	—
Side scrapers	15	4	—	—	4	19.1	26.7
Combination notch/borer/ denticulates	7	1	—	—	1	4.8	14.3
Bend-break tools	3	1	—	—	1	4.8	33.3
Backed and snapped Unifaces	6	2	—	—	2	9.5	33.3
Piercers	18	4	—	—	4	19.1	22.2
Denticulates	3	—	—	—	—	—	—
Retouched flakes	6	1	—	—	1	4.8	16.7
Other unifaces	7	2	—	—	2	9.5	28.6
TOTALS	159	21	0	0	21	100.0	13.2

*C = Collingwood; B = Bayport; O = Other.

centration. Either of these two interpretations being the case, the areal size estimate of A-northeast would be reduced to around 90 square meters.

A-East

This area covers about 80 square meters and appears to have been almost entirely exposed by excavation. There are a few disturbances such as roots in the area. Despite the presence of one feature (Feature 4) associated with later prehistoric occupations at the site, later occupation debris is rare.

Only one probable feature (Feature 10) associated with the Paleo-Indian occupation of the site occurs in A-east. This was located on the northern periphery of the area. Twenty-one tools were recovered from A-east, excluding a heat-fractured Bayport point base fragment which fits together with a fragment from A-northeast. As noted above, this inter-area link probably resulted from plowing. Since the Bayport chert is very strongly associated with A-northeast, we assume that this point fragment was originally from that area. Excluding this fragment, all of the tools from A-east are on Collingwood chert.

No one tool class or type strongly predominates in A-east as was the case for fluted bifaces in A-northeast. In fact, fluted bifaces are rare. Excluding the Bayport fragment noted above and a small midsection fragment

TABLE 54
Artifact Total, A-West

Type	Site Total	C	B	O	Area Total	% of Area Total	% of Site Total
Fluted points	12	3	—	—	3	12.5	25.0
Fluted preforms	12	2	—	—	2	8.3	16.7
Fluted bifaces	8	4	—	—	4	16.7	50.0
Unfluted preforms	3	—	—	—	—	—	—
Alternately beveled bifaces	5	—	1	—	1	4.2	20.0
Concave-base bifaces	2	1	—	—	1	4.2	50.0
Oval bifaces	2	—	—	—	—	—	—
Channel flake point	1	—	—	—	—	—	—
Pièce esquillée (?)	1	—	—	—	—	—	—
Backed biface	1	1	—	—	1	4.2	100.0
Triangular end scrapers	10	1	—	—	1	4.2	10.0
Offset end scrapers	6	—	—	—	—	—	—
Fluted end scrapers	2	—	—	—	—	—	—
Narrow end scrapers	5	2	—	—	2	8.3	40.0
Combination narrow/wide end scrapers	3	2	—	—	2	8.3	66.7
Proximal end and side scrapers	8	—	—	—	—	—	—
End and concave side	2	—	—	—	—	—	—
End and convex side	2	1	—	—	1	4.2	50.0
Other end scrapers	9	1	—	—	1	4.2	11.1
Side scrapers	15	3	—	—	3	12.5	20.0
Combination notch/borer/ denticulates	7	—	—	—	—	—	—
Bend-break tools	3	—	—	—	—	—	—
Backed and snapped unifaces	6	1	—	—	1	4.2	16.7
Piercers	18	—	—	—	—	—	—
Denticulates	3	—	—	—	—	—	—
Retouched flakes	6	1	—	—	1	4.2	16.7
Other unifaces	7	—	—	—	—	—	—
TOTALS	159	23	1	0	24	100.2	15.1

C = Collingwood; B = Bayport; O = Other.

of a biface which cannot be assigned to the point or preform type, the only fluted biface is a preform. This item is the small atypical artifact (Fig. 22i) which we suggested earlier might be the product of a novice knapper. The only unifacial tool which can be said to be common in the area is piercers or spurs. These include four examples which constitute 22.2% of the site total and 19.9% of the assemblage at A-east.

Flaking debris is rare at A-east (around 4 flakes per two-meter unit). The most recovered in an excavation unit was 12 items from the square containing Feature 10. This total is inflated because of the excavation techniques employed in the subsoil. The low frequency of waste is undoubtedly related to the general lack of biface-associated activity in the area. The emphasis on uniface-related activities is evident not only in the tools recovered but also by the fact that uniface retouch flakes were recovered in almost equal frequency to biface flaking debris (Table 60). Finally, it can be noted that the flaking debris is consistent with the tool assemblage in that Bayport debris is rare (1 flake) and that no channel flakes were recovered (Table 61).

A-West

This site area covers 88 square meters. The area was highly disturbed by root action and perhaps by later prehistoric occupations. The concentration partially co-

TABLE 55
Artifact Total, A-Centre

Type	Site Total	C	B	O	Area Total	% of Area Total	% of Site Total
Fluted points	12	—	—	—	—	—	—
Fluted preforms	12	—	—	—	—	—	—
Fluted bifaces	8	1	—	—	1	2.2	12.5
Unfluted preforms	3	—	—	—	—	—	—
Alternately beveled bifaces	5	1	1	—	2	4.4	40.0
Concave-base	2	—	—	—	—	—	—
Oval bifaces	2	1	—	—	1	2.2	50.0
Channel flake point	1	—	—	—	—	—	—
Pièce esquillée (?)	1	—	—	—	—	—	—
Backed biface	1	—	—	—	—	—	—
Triangular end scrapers	10	1	1	—	2	4.4	20.0
Offset end scrapers	6	1	—	—	1	2.2	16.7
Fluted end scrapers	2	1	—	—	1	2.2	50.0
Narrow end scrapers	5	—	—	—	—	—	—
Combination narrow/ wide end scrapers	3	—	—	—	—	—	—
Proximal end and side scrapers	8	3	1	—	4	8.9	50.0
End and concave side	2	1	—	—	1	2.2	50.0
End and convex side	2	—	—	—	—	—	—
Other end scrapers	9	2	—	—	2	4.4	22.2
Side scrapers	15	6	—	1	7	15.6	46.7
Combination notch/ borer/denticulates	7	2	—	—	2	4.4	28.6
Bend-break tools	3	—	—	—	—	—	—
Backed and snapped unifaces	6	1	—	—	1	2.2	16.7
Piercers	18	12	—	1	13	28.9	72.2
Denticulates	3	2	—	—	2	4.4	66.7
Retouched flakes	6	2	—	—	2	4.4	33.3
Other unifaces	7	3	—	—	3	6.7	42.9
TOTALS	159	40	3	2	45	99.7	28.3

C = Collingwood; B = Bayport; O = Other.

TABLE 56
Artifact Total, A-Southeast

Type	Site Total	C	B	O	Area Total	% of Area Total	% of Site Total
Triangular end scrapers	10	1	—	1	2	22.2	20.0
Proximal end and side scrapers	8	3	—	—	3	33.3	37.5
End and concave side	2	1	—	—	1	11.1	50.0
Other end scrapers	9	1	—	—	1	11.1	11.1
Side scrapers	15	1	—	—	1	11.1	6.7
Retouched flakes	6	1	—	—	1	11.1	16.7
TOTALS	159	8	—	1	9	99.9	5.7

C = Collingwood; B = Bayport; O = Other.

TABLE 57
Artifact Totals, Grid B

Type	Site Total	C	B	O	Area Total	% of Area Total	% of Site Total
Fluted preforms	12	1	—	—	1	16.7	8.3
Unfluted preforms	3	1	—	—	1	16.7	33.3
Triangular end scrapers	10	1	—	—	1	16.7	10.0
Offset end scrapers	6	1	—	—	1	16.7	16.7
Narrow end scrapers	5	1	—	—	1	16.7	20.0
Combination notch/ borers/denticulates	7	1	—	—	1	16.7	14.3
TOTALS	159	6	—	—	6	100.2	3.8

C = Collingwood; B = Bayport; O = Other.

TABLE 58
Artifacts Unassignable to Areas

Type	Site Total	C	B	O	Area Total	% of Area Total	% of Site Total
Alternately beveled bifaces	5	—	1	—	1	5.9	20.0
Triangular end scrapers	10	1	—	—	1	5.9	10.0
Offset end scrapers	6	3	—	—	3	17.7	50.0
End and convex side	2	—	1	—	1	5.9	50.0
Other end scrapers	9	3	1	—	4	23.5	44.4
Combination notch/borer/ denticulates	7	1	—	—	1	5.9	14.3
Bend-break tools	3	1	—	—	1	5.9	33.3
Backed and snapped unifaces	6	2	—	—	2	11.8	33.3
Denticulates	3	1	—	—	1	5.9	33.3
Retouched flakes	6	1	—	—	1	5.9	16.7
Other unifaces	7	1	—	—	1	5.9	14.3
TOTALS	159	14	3	—	17	100.2	10.7

C = Collingwood; B = Bayport; O = Other.

TABLE 59
Waste Flakes by Site Area

Area	Bayport		Collingwood		Total	
	R*	X*	R	X	R	X
A-West	0–2	0.5	1–106	23.3	1–107	23.8
A-Centre	0–1	0.02	0–46	6.8	0–46	6.8
A-East	0–1	0.05	0–12	3.7	0–12	3.8
A-Northeast	0–6	1.4	0–84	7.8	0–84	9.2

*Range and mean per two-meter excavation unit assuming no cross-mends between separate units.

incides with a dense cluster of later lithic debris and features (Figs. 65, 82: Concentration 1). As noted in Chapter 6, there is a dense concentration of Paleo-Indian tools and waste (Figs. 72, 76) located almost exactly in the center of the excavated area which might be a feature remnant.

Excluding a few biface fragments on Bayport chert (only one of which is definitely Paleo-Indian), the A-west assemblage is primarily Collingwood chert (Table 54; Fig. 83). The area has a high ratio of bifaces to unifaces (Table 60), being exceeded only by A-northeast which, as noted above, contains numerous bifaces from a cache. The majority of the bifaces are fluted (25% of the site total; Table 54) and includes both points and preforms. The former are represented by a tip-impacted

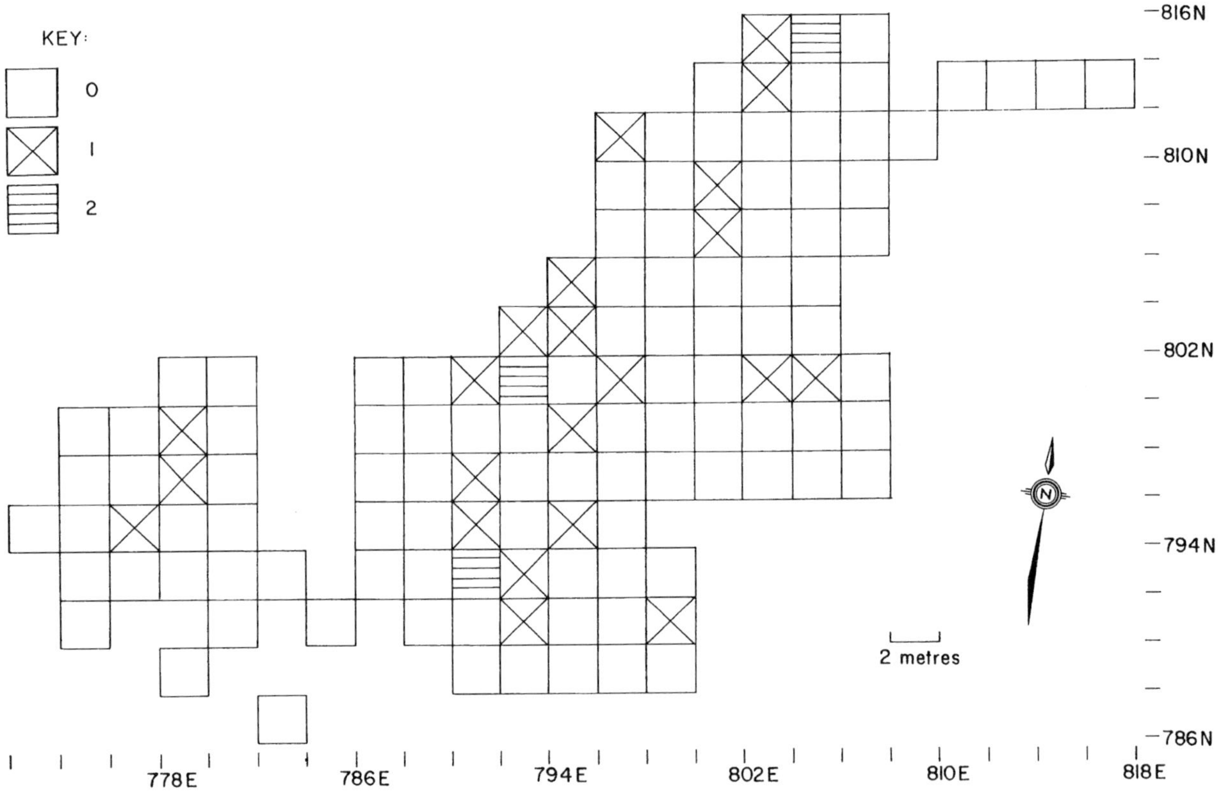

Figure 77. Normal and end biface thinning flake density.

point with a broken ear and part of a base (Fig. 20e, g). As with the broken basal ends at A-northeast, this suggests rehafting of new points to replace the use-damaged specimens. Both of the preforms from A-west were broken in manufacture and along with numerous channel flakes, indicate the presence of point manufacture.

The majority of the unifacial tool types and classes are represented by a few items. However, piercers are absent in contrast to their representation in all other site areas. Tools incorporating narrow end scraper bits are relatively common (50% of the site total) in contrast to the other site areas.

A-west includes a relatively large and dense amount of lithic waste (Table 59; Fig. 76). There is also a high debris-to-tool ratio (Table 60) as is to be expected given the number of channel flakes and the emphasis on biface reduction and use evident in the tool assemblage. However, the absolute number of uniface retouch flakes, which is the highest for any site area, also suggests uniface reduction was important.

Comparisons: Peripheral Clusters

Interpretation of the meaning of the peripheral clusters and of the similarities and contrasts among them and with A-centre (see below) is difficult. Here, we will focus on delineating patterns in the data and with suggesting possible explanations. These explanations assume that the peripheral areas (as well as A-centre) were occupied contemporaneously.

As noted above, the few cross-mends between the major clusters may be a product of plowing and therefore, cannot be used as definitive evidence of contemporaneity. The cache in A-northeast might suggest anticipation of return to the site. Hence, several occupations may be a more likely scenario. However, this cache was not retrieved. The possibility also exists that a cache could be retrieved in subsequent visits to the site *area* without necessarily occupying the site *per se*. At the present time, we favor a simultaneous occupation of the areas for two main reasons. First, and accepting some distortion due to plowing, the spatial separation be-

TABLE 60
Debris:Tool Ratios and Biface:Uniface Ratios*

Ratio	A-West	A-Centre	A-East	A-Northeast
Collingwood debris/tool				
with tool fragments	10.4:1	4.1:1	2.9:1	6.3:1
without tool fragments	17.3:1	7.4:1	3.5:1	9.9:1
Bayport debris/tool			—	2.5:1
with tool fragments	3.0:1	0.3:1		
without tool fragments	9.0:1	0.3:1	—	3.3:1
Total debris/tool	9.6:1	3.9:1	2.9:1	5.5:1
with tool fragments				
without tool fragments	16.9:1	6.9:1	3.6:1	7.2:1
Collingwood biface to uniface debris ratios	2.3:1	1.8:1	1.1:1	7.1:1
Bayport biface to uniface debris ratios	2.0:1	—	—	0.5:1
Total biface to uniface debris ratios	2.3:1	1.8:1	1.1:1	3.4:1
Collingwood biface to uniface tool ratio	.67:1	.06:1	.17:1	1.6:1
Bayport biface to uniface tool ratio	—	.50:1	—	5.0:1
Total biface to uniface tool ratio	.75:1	.08:1	.17:1	2.2:1

*Based on totals of flakes and tools after refitting; excludes tools outside the excavated area and tools on unknown materials.

tween adjacent clusters is quite regular (2–4 m). Second, there is a symmetrical overall patterned layout with the peripheral clusters being arranged in a semicircle around A-centre.

In addition to their positions at the margins of the main site area, the peripheral clusters have certain commonalities. First, all the excavated clusters have a similar areal extent (80–100 square meters, assuming the Feature 13 flake cluster is not associated with A-northeast; see arguments above). Surface finds suggest that the unexcavated clusters (A-southeast and B) are of a comparable size. Second, and excluding the probable 13 points from a cache in A-northeast, a similar number of worn-out tools and preforms was recovered from each (20–25 items). Estimates of the yield of the unexcavated clusters are difficult to derive. However, the low number of finds from each separate surface examination of these areas (usually 0–2 items) suggests that they have low overall yields like the excavated clusters. Finally, the excavated clusters all provide some evidence of the presence of one or possibly two "artifact" features lacking visible outlines in the soil.

There is some suggestion that different individuals were present in each of the excavated peripheral clusters. The best evidence of this is the major clustering of Bayport chert in A-northeast. This strongly suggests that an individual(s) was present at the locus who, in contrast to the individuals responsible for the other clusters, had a means of access to that chert so that it could be obtained in quantity.

In addition to the distributional pattern of Bayport chert, there are other variations between the marginal clusters which are suggestive of the presence of different individuals. Although hampered by small sample sizes, the best evidence is to be found in the fluted biface assemblage from the various clusters. There is formal and technological variation among the fluted bifaces which correlates well with the excavated areas. This variation is not due to factors such as resharpening or different raw materials. Thus, it might be indicative of different motor habits or degrees of flint-knapping skill among individuals (presumably males) at the various loci.

To begin, the A-west assemblage of these bifaces exhibits several distinctive characteristics. First, all three items with markedly off-center flutes are from A-west. Furthermore, in all cases, although the base of the flute is relatively well-centered, the flute runs diagonally toward the left lateral edge. These items include: one finished point with off-center flutes on both faces; one finished point with such a flute on one face; and one preform with the off-center flute on one face (e.g. Fig. 28).

Second, the only two points broken in fluting by plunging flutes are from this site area. This may be related to the off-center flutes since one preform probably broke when the flute encountered a lateral edge (Fig. 28). However, the other biface broken by this form of termination (Fig. 24a) had a well-centered flute. Overall, and especially in relation to the fluted bifaces from A-northeast, one gets the impression of a less-skilled point-maker at A-west. This is evident not only in the off-center fluting and the presence of plunging flutes but also, by the presence of the single finished point with very short fluting on both faces (Fig. 20e).

Finally, the Barnes basal finishing technique removals on points from the site can either feather out and blend with the main flute scars they overlie or they can terminate abruptly in a hinge or step termination. On all three points from A-west where this trait is observable, the basal finishing flakes have feathered termination and are difficult to distinguish from the main flute scar.

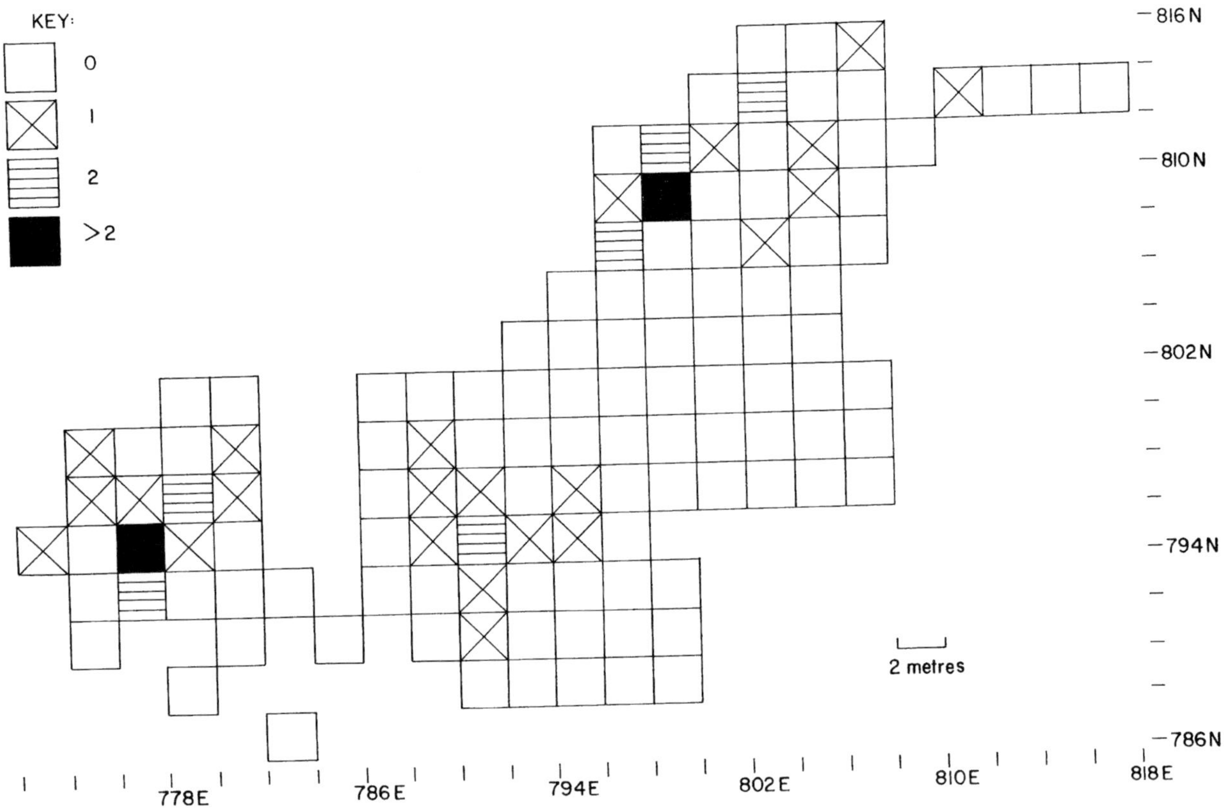

Figure 78. Distribution of channel flake fragments.

The A-northeast fluted biface sample is distinctive in several respects. First, and by extension from previous discussions, it includes the only Bayport chert points and preforms from the site. Second, it accounts for the only examples of Barnes basal finishing flakes that terminate abruptly in hinge or step terminations (8 of 9 items on which this trait is observable; e.g., Figs. 20a-c, 21, 22d, 23a). These occur whether or not the points are of Bayport chert, the soft Collingwood chert or the hard, banded Collingwood chert. Third, A-northeast includes the only multiple fluted points (3 examples) and preform (1 example) from the site. In all cases, the main flute is well-centered and the second flute is a long, narrow, linear one (e.g. Fig. 22b). Finally, all points and most preforms have had pronounced, linear flakes removed along the base of the flute's lateral edges. This trait is lacking on the other site points.

Area A-east exhibits little evidence of point manufacture. No channel flakes were recovered and only small amounts of biface debris were noted. This may be due solely to activity differences between areas. However, this contrast could be due as well to the skill of the knapper(s) present at this locus relative to the other areas. The only preform from this area is the small atypical one (Fig. 22i) which we suggested earlier was the product of a novice knapper. In short, if an unskilled knapper had been present at this locus, he may not have produced much in the way of points.

In summary, certain contrasts between the excavated peripheral clusters suggest that different individuals were present at each loci. Presumably, these individuals were males as the main contrasts involve the fluted biface assemblages from each area. Taking this inference a step further, one can speculate that these clusters represent social differences as well. The presence of artifacts in these clusters such as the channel flake point from A-northeast and the trianguloid end scrapers found in most excavated and unexcavated marginal clusters can be used to suggest that the individuals at each area did not solely include adult males. As noted

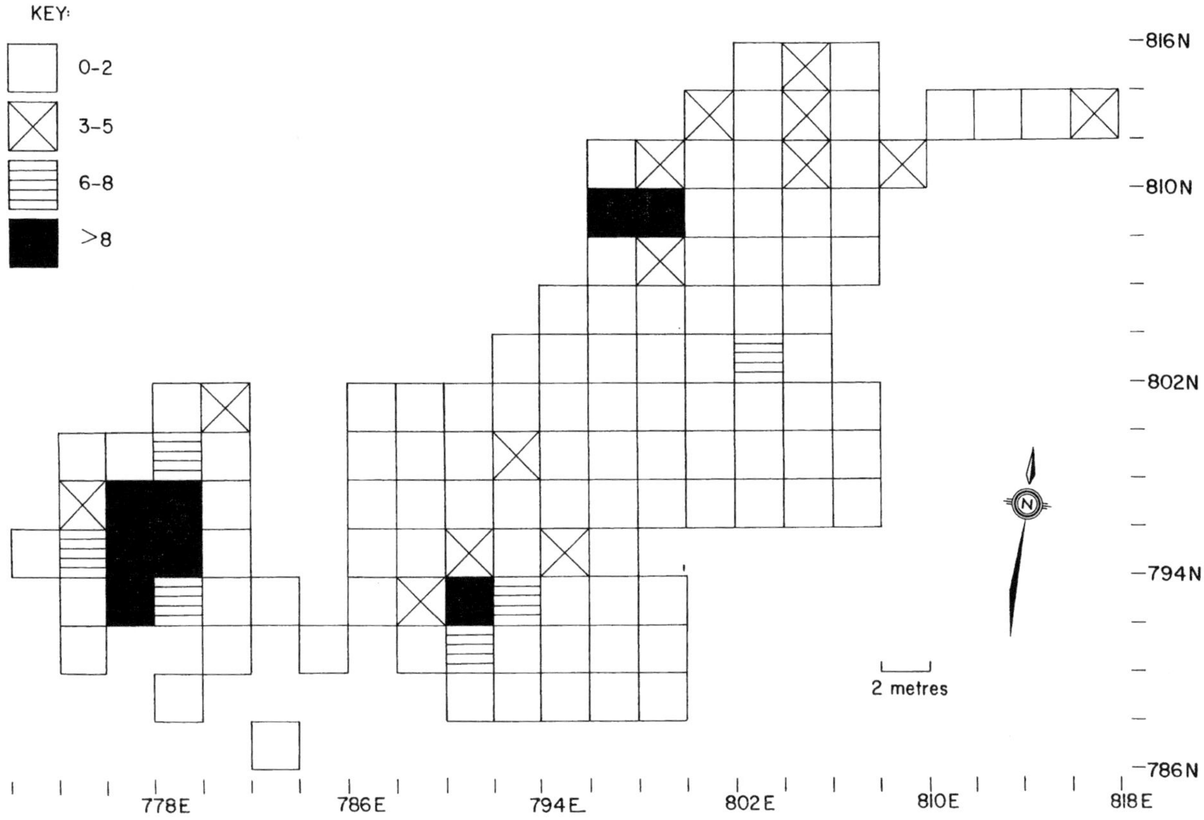

Figure 79. Distribution of other biface reduction flakes.

earlier, channel flake points may be toys. Moreover, at least some data such as the Crowfield site presumed burial cache suggest trianguloid end scrapers were used by females (see Deller and Ellis 1984: 51–52). On these bases, and compared with the small areal extent of each cluster, it seems possible that each might represent the location of small social units such as minimal family groups.

In addition to the suggestion of differences in individuals present at each marginal cluster, there is evidence that the nature of the activities in each cluster varied. This is based on the differential distribution of certain artifact types and classes. Points and preforms are a good example. Their manufacture was important at A-west and A-northeast but less important at A-east. Assuming point basal sections broken in use are indicative of rehafting (as well as hunting activities), then this activity occurred both in A-northeast and A-west. It is interesting to note as well that tools incorporating narrow end scraper bits primarily came from these two excavated areas. This might suggest a functional association of these tools with points, perhaps in the making of hafts. This is largely speculation but it is worth searching for a similar pattern of association on other sites.

Other suggestions of activity differences involve the distribution of piercers, side scrapers and proximal end and side scrapers. The piercers are rare or absent in A-northeast and A-west while they account for 19% of the A-east artifacts (4 examples). However, this difference need not mean that there were great differences in the time devoted to activities involving piercers. These are a tool form which can be rapidly and easily made and which, because of their delicate nature, probably had short use-lives relative to most bifaces and unifaces (see Curran 1984:17).

Side scrapers as a class probably include much use-variation. Nonetheless, their total absence from areas such as A-northeast but the presence of three or four examples in the other excavated marginal areas is sug-

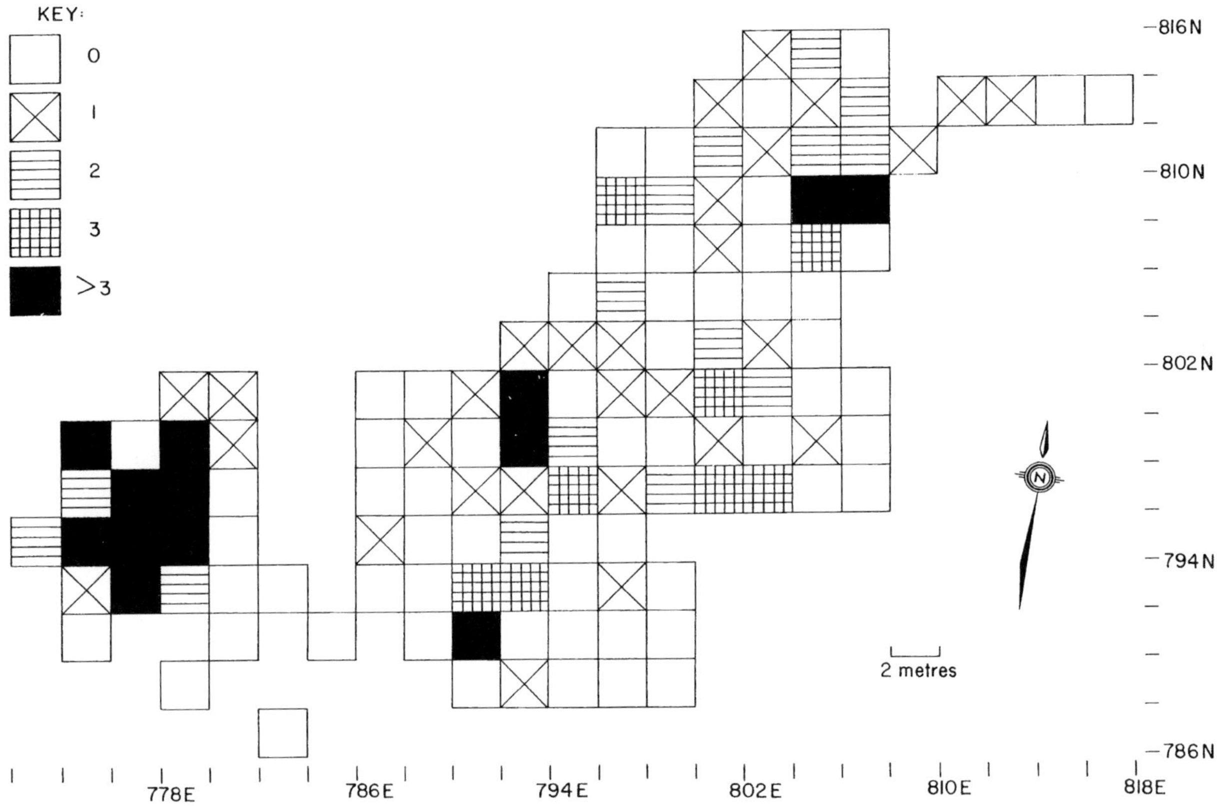

Figure 80. Distribution of uniface retouch flakes.

gestive of activity differences. Finally, proximal end and side scrapers are rare in the excavated clusters of concern here. The small number of surface finds from A-southeast includes three of these tools (Table 56).

Overall, the suggestion is that the activity differences between the marginal site areas were minor. In sum, there is no evidence that the various site areas represent specialized activity areas. Fluted biface production and use is a possible exception but, if A-east represents a novice knapper as suggested above, these differences can be largely accounted for by this factor. Excluding fluted biface production and use, the other minor contrasts between areas could represent simply a short duration of occupation. It is possible that there will be less diversity of tool types and classes between site areas as occupation span increases. In other words, the longer the site is occupied, the greater will be the probability that each individual or group of individuals present at a locus will carry out a larger portion of the total set of activities (see Curran 1984:16). In the case of short-term occupations, one might see a "founder effect" wherein variability simply indicates differences in the state of an individual's tool kit or equipment as they arrived at a site. We believe that this factor could easily account for inter-area differences in the use of certain tools such as narrow end scrapers, piercers, side scrapers and proximal end and side scrapers. The very low yield of artifacts versus other sites in the east such as Debert (MacDonald 1968) and Vail (Gramly 1982) is certainly suggestive of a relatively short-term occupation.

Alternatively, the differences between the peripheral clusters may not be due to a short occupation of the site as a whole but to a short period of use of these areas for lithic-related activities. If the peripheral clusters each represent different social units and A-centre represents a communal work area (see below) then it is possible that each social unit may have carried out some of its activities in the central area rather than in their respective peripheral clusters. These activities could potentially vary between social units in terms of their being

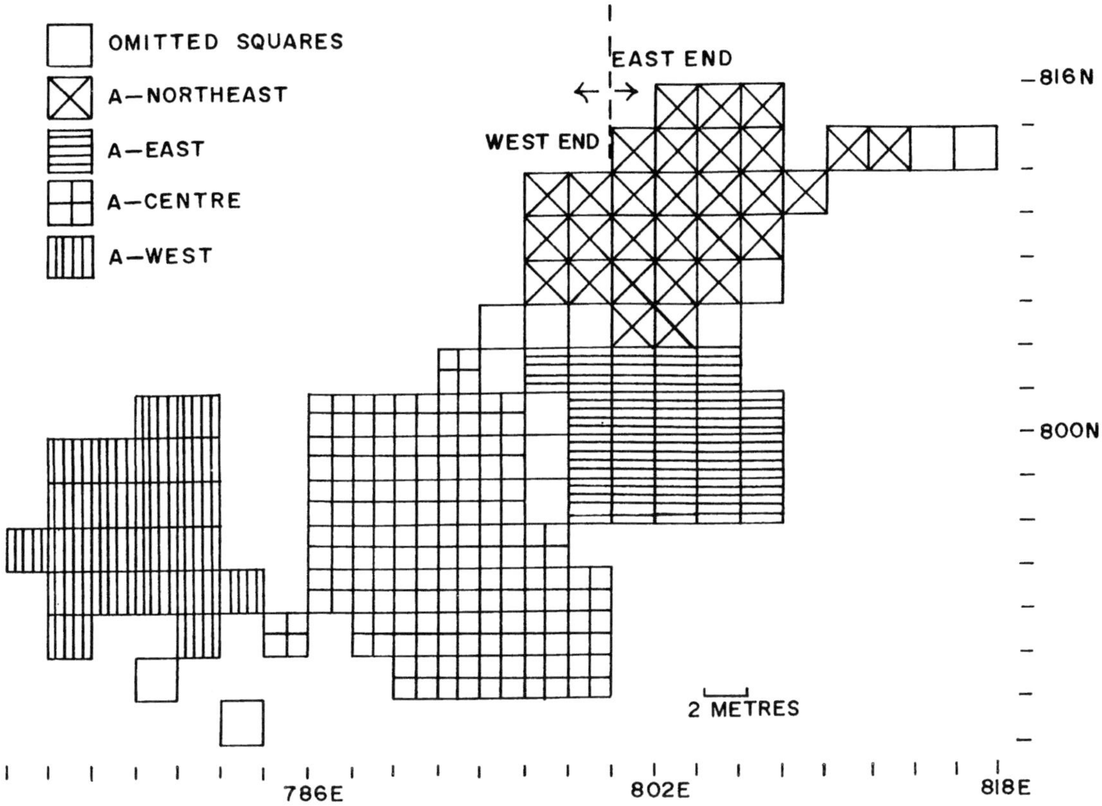

Figure 81. Flaking debris assignments by excavated unit.

carried out in a peripheral cluster or in the central area itself. Hence, even if the tool kit of each social unit was in a similar state, variability between the peripheral clusters could be due simply to variation in the location where an activity was carried out. If the tool kits were in different states of disrepair or preparation, combined with this factor activity variation between the peripheral clusters would be even greater.

Central Cluster

The central cluster, A-centre, is significantly different from the peripheral clusters. First, its two suspected Paleo-Indian features (Features 3 and 8) had visible outlines. This suggests their function was different from that of the artifact features. Second, roughly twice as many tools (Table 55) were recovered. Almost all of these were on Collingwood chert. Finally, the area covers at least 164 square meters and may extend further to the south. This is almost twice the size of the peripheral clusters.

These latter two contrasts assume A-centre is one homogeneous area. There is some evidence of two subclusters centering on each of the features in the north and south halves. This is suggested by density diagrams of tool fragments (Fig. 73), total flaking debris (Fig. 76) and some specific types of debris (Figs. 77, 80). Combined with the fact the area is roughly twice as large and contains twice as many tools as the marginal areas, this might suggest A-centre consists of two discrete clusters, each of which is comparable to those in the surrounding areas. However, there is some evidence that this is not the case.

First, even though roughly twice as many tools were recovered, a high absolute and proportional number of these are piercers (Table 55; Fig. 87). As noted earlier, these can be rapidly and easily made and undoubtedly had short use-lives. Therefore, the number of tools is inflated in comparison to the other clusters where these tools are rare. Moreover, the tool assemblage as a whole is not evenly distributed at the north and south ends so even if each is separate, tool frequencies would vary somewhat from those in other areas.

Second, there is little suggestion among the refits of

the separation of A-centre (Fig. 74). However, the fact the site has been plowed on a north-south axis may be having an effect here.

Third, density diagrams of at least the channel flakes (Fig. 78) do not cluster in the two feature areas. In fact, these flakes are largely restricted to an area between the two features. This suggests that the channel flakes are associated with both features and that the sub-clusters are closely related.

Finally, these sub-clusters are more similar to each other in terms of activity indicators than they are to the peripheral clusters. This again suggests a close association. As noted above, the suspected features associated with each sub-cluster had visible outlines. Moreover, the only tools recovered from each feature are piercers. Feature 3 included one piercer while Feature 8 had three if we include the uniface with a small spur (Fig. 62c). In addition, the features are similar in that they are both centers for concentrations of biface thinning flakes and uniface retouch flakes (Figs. 78, 81).

Tool forms also suggest activity similarities. Although it is difficult to separate precisely the sub-clusters, based on the density diagrams of tools and waste flakes, they are roughly divided along the 795 to 796N line. On this basis, both sub-clusters include items which are rare in one or more of the peripheral clusters. For example, piercers, including the uniface with spur (Fig. 62c) from Feature 8 make up more by percentage (greater than 20%) in both ends than they do in any other cluster. Moreover, both ends clearly include proximal end and side scrapers, trianguloid end scrapers and simple side scrapers which can be lacking or rare in the peripheral clusters. In fact, both ends have the same number of these first two tool forms (Fig. 86).

At the opposite extreme, both A-centre sub-clusters are also the most similar to one another in that they can lack items very common in certain peripheral clusters. For instance, both sub-clusters generally lack narrow end scrapers (Fig. 86) and fluted bifaces (Fig. 85) which are relatively common in areas such as A-northeast and A-west. The presence of channel flakes from A-centre indicates fluted point manufacture occurred (Table 61) despite the general absence of fluted bifaces. The major difference appears to be the lack of points broken in use which are presumably indicative of point re-hafting activities. This factor may partially account for the relatively low percentage of bifaces at A-centre despite relatively high percentages of biface debris and debris-to-tool ratios (Table 60).

On the above bases then, we feel justified in suggesting A-centre is one integral unit as opposed to the surrounding clusters. The two sub-clusters are taken to indicate simply that activities tended to focus around the two probable features in the area. The larger size and larger artifact yield in A-centre suggests it represents something different from the peripheral clusters. Given that the peripheral areas represent the location of different social units, we would speculate that A-centre represents a communal work area shared and jointly used by the individuals represented in those peripheral areas. This interpretation is suggested by three sources of data.

The first is a criterion of accessibility to all site occupants. A-centre, by definition, is central to the site. As such, it would be easily and conveniently accessible for use by all site occupants.

Second, a communal area would be used by more individuals; therefore, we should expect it to be larger in areal extent.

Finally, we argued that the activity contrasts between the peripheral clusters may be at least partially a product of a short duration of occupation and differences in the state of an individual's tool kit on arrival at the site. A communal area, used by all site occupants with tool kits in various states of disrepair, should be expected to exhibit a wider overall range of activities. Alternatively, if there was variation in the activities that each social unit carried out in A-centre, either alone or in combination with variation in the state of different tool kits, this should result in A-centre witnessing a wider range of activities.

Almost all activities suggested by well-represented tool forms of use-significance are represented in A-centre. The only exceptions are points damaged in use and narrow end scrapers, both of which might be related to point hafting. Tool forms well-represented in A-centre but which can be rare or lacking in one or more of the peripheral clusters include piercers, proximal end and side scrapers, trianguloid end scrapers and the general class of side scrapers. Moreover, even rare tool forms on the site (less than 6 items) are almost all represented in A-centre but are often absent from the other clusters. This includes certain highly distinctive types of probable use-significance such as: alternately beveled bifaces (2 of 4 assignable to clusters), small oval bifaces (1 of 2), backed and snapped unifaces (1 of 4), offset end scrapers (1 of 3) and end and concave scrapers (1 of 2). As well, more general classes of tools, which probably measure use-variability at a gross level, are also consistently found in A-centre. These consist of denticulates (2 of 2), retouched flakes (2 of 5) and combination notch/borers/ denticulates (2 of 6).

Overall, and if we treat fluted bifaces and tools with narrow end scraper bits each as single classes and ig-

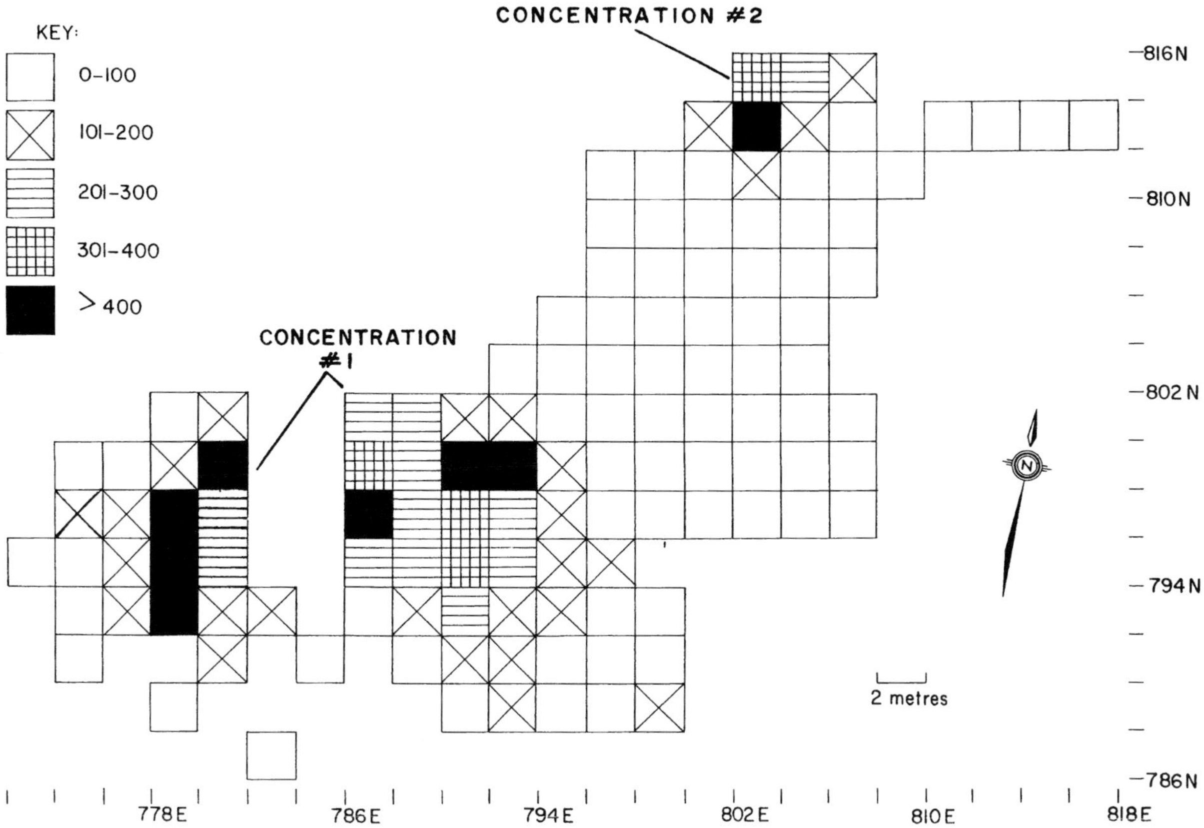

Figure 82. Distribution of non-Paleo-Indian flaking debris.

TABLE 61
Flaking Debris by Type and Area*

Area	Channel Flakes	Biface Thinning Flakes	Other Biface Flakes	Uniface Debris	Flat & Other Flakes	Totals
West	13	3	125	62	151	354
Centre	9	12	70	52	133	276
East	0	2	20	20	33	75
Northeast	17	7	99	36	107	266
TOTALS	39	24	314	170	424	971

*Includes totals for Collingwood and Bayport chert debris after cross-mends; excludes surface finds and peripheral squares not easily assigned to specific site areas.

nore the "other end scraper" and "other unifacial tool" classes, some 66.7% of the tool forms are present in A-centre. All other excavated clusters contain less than 52.4% of the tool forms. Moreover, it is notable that the tool forms not represented in A-centre are, excluding narrow end scrapers, very rare forms represented by only one (a channel flake point, a backed biface and a doubtful *pièce esquillée*) or two (concave-base bifaces; bend-break tools) items.

The impression, therefore, is that overall a wider range of activities occurred in A-centre than in any of the individual peripheral clusters. Phrased another way, A-centre has a richer (see Rice 1981) assemblage than the other clusters. This does not mean that certain activities were not more important than others in the central cluster. If we focus on the frequency of artifact types in the area instead of largely presence/absence, there is a suggestion that at least one activity was a major focus. This included the unknown activity which involved piercer use. The large number of these tools from the central area (over 70% of the site total) are, from this perspective, suggestive of some kind of specialized activity. Of course, the short use-life of these tools may mean they are over-emphasized. However,

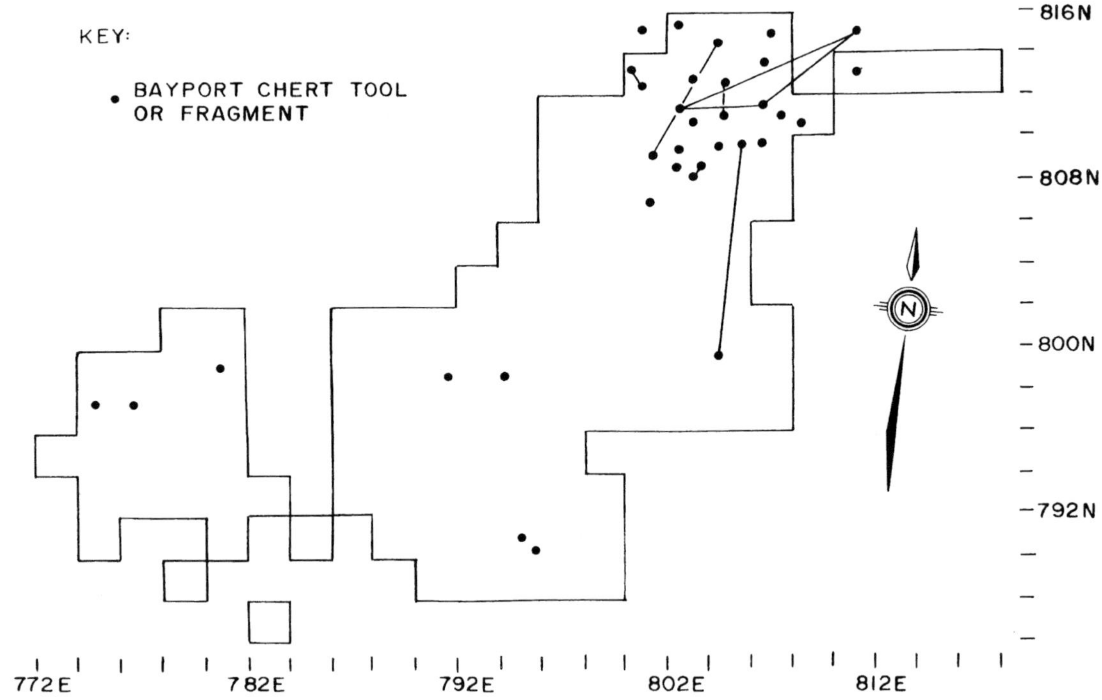

Figure 83. Distribution of Bayport chert tools and fragments.

TABLE 62
Frequency of Piercers, Parkhill Industry Sites

Site or Site Area	Total Sample Size	Piercers N	Piercers %*
Thedford II: A-West	24	0	0.0
Thedford II: A-Centre	45	13	28.9
Thedford II: A-East	21	4	19.1
Thedford II: A-Northeast with cache	37	1	2.7
Thedford II: A-Northeast without cache	24	1	4.2
McLeod	18	1	5.6
Barnes	59	0	0.0
Banting Area A-West	46	7	15.2
Parkhill: Grid B	94	0	0.0
Parkhill: Grid C	37	5	13.5
Parkhill: Grid D	56	2	3.6

SOURCES: Ellis 1984 (McLeod site); Voss 1977: Table 2 (Banting site); Storck 1979: Table 2 (Banting site, excludes cores and fragments); Roosa and Deller 1982:12 (Parkhill site, excludes "fluted drill").
*Percentage of total site assemblage or site area.

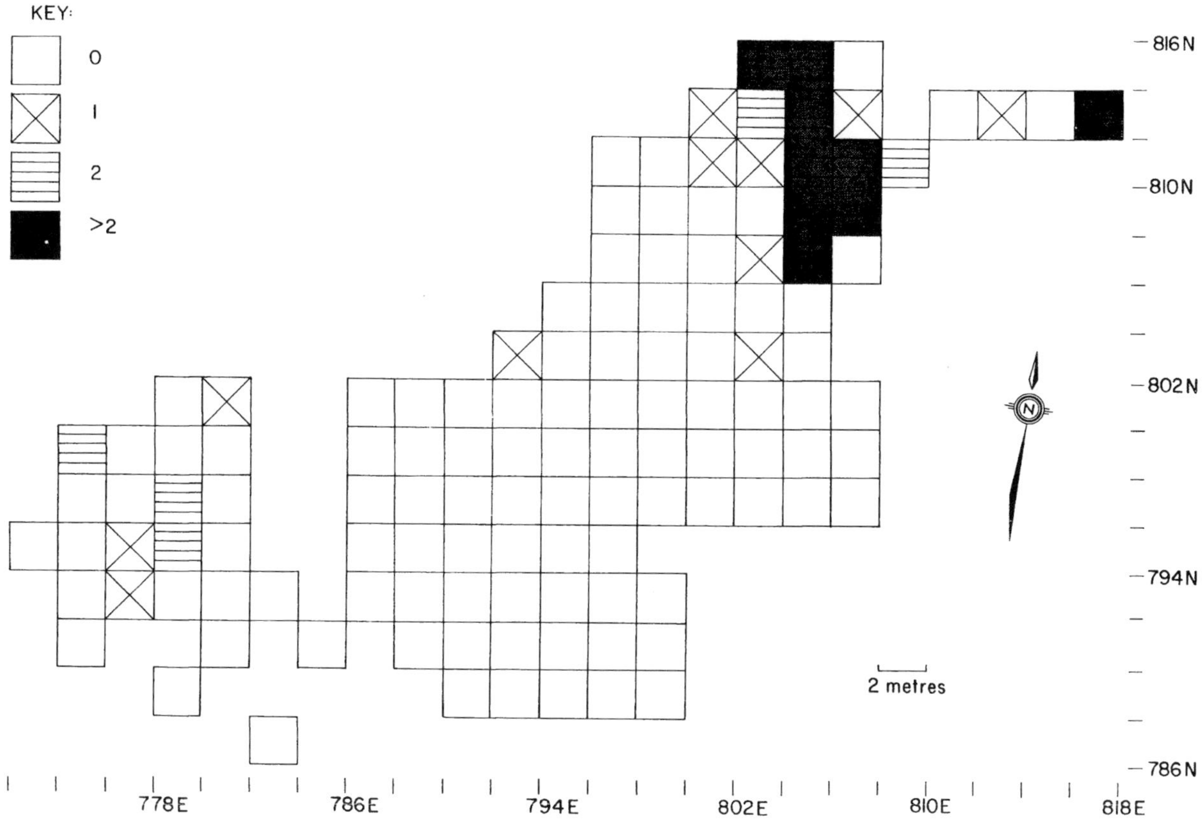

Figure 84. Distribution of Bayport chert flaking debris.

tools may mean they are over-emphasized. However, the very fact that the percentage of these tools is higher than that reported from any other Parkhill industry assemblage (Table 62) suggests a specialized activity involving these tools. This once again suggests A-centre is different from the peripheral clusters.

Summary

Excavations and surface collections at Thedford II suggest the presence of six discrete artifact clusters. Five of these, located around the periphery of the main site area, are similar in size, artifact yield and suspected features. Based on variability in raw materials and in fluted point production and morphology, this suggests that different individuals were present at these loci and that each represents the location of a small social unit such as a family group. There are suggestions of activity differences between these clusters and we argue that these are a product of the skill of the point-makers at each locus and a short duration of occupation or a limited time of use for lithic-related activities.

The remaining cluster differs from the peripheral clusters in terms of being larger, yielding more tools, and having indications of different kinds of features. As a result of its central location, larger size and indications that a wider range of activities took place in this cluster, we suggest that it represents a communal work area, employed by all the site inhabitants. The overall layout suggests a series of family groups arranged in a semicircle around the central work area. This is a similar interpretation to that offered by Fitting et al. (1966) for the Holcombe site in Michigan.

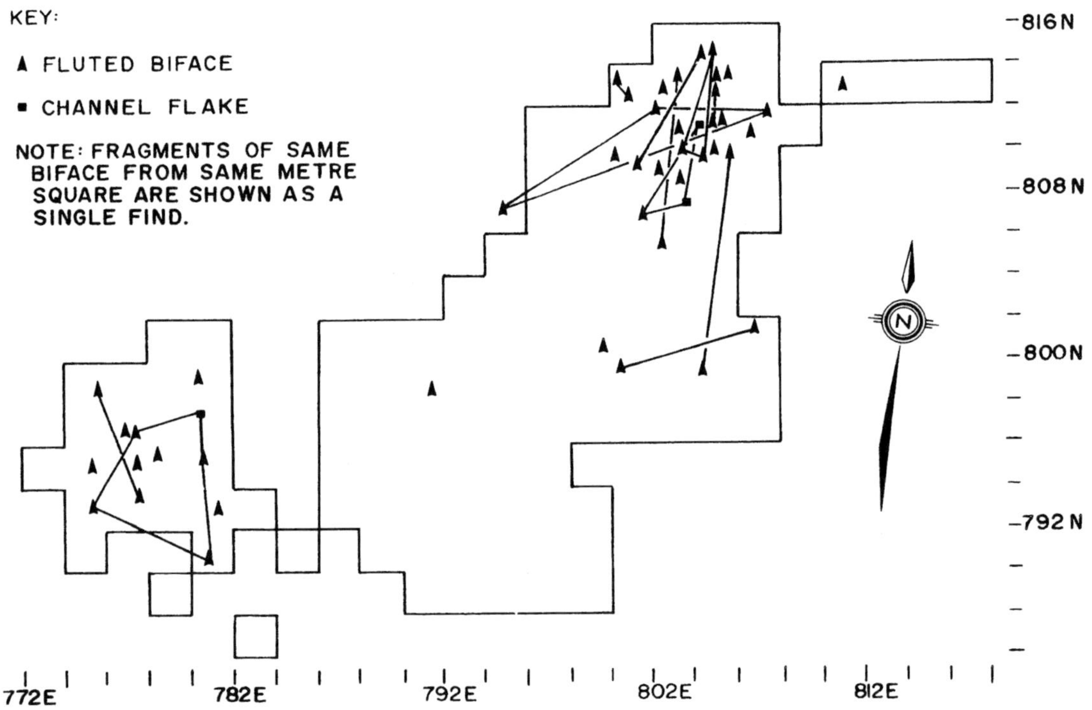

Figure 85. Distribution of fluted bifaces and matching channel flakes.

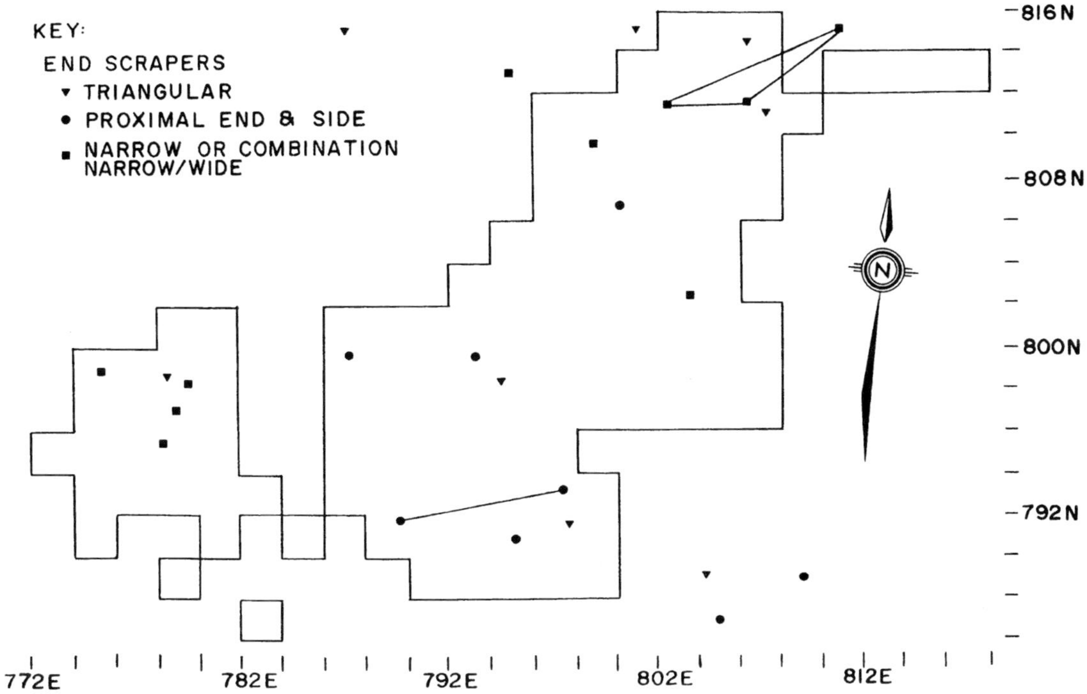

Figure 86. Distribution of certain end scraper types.

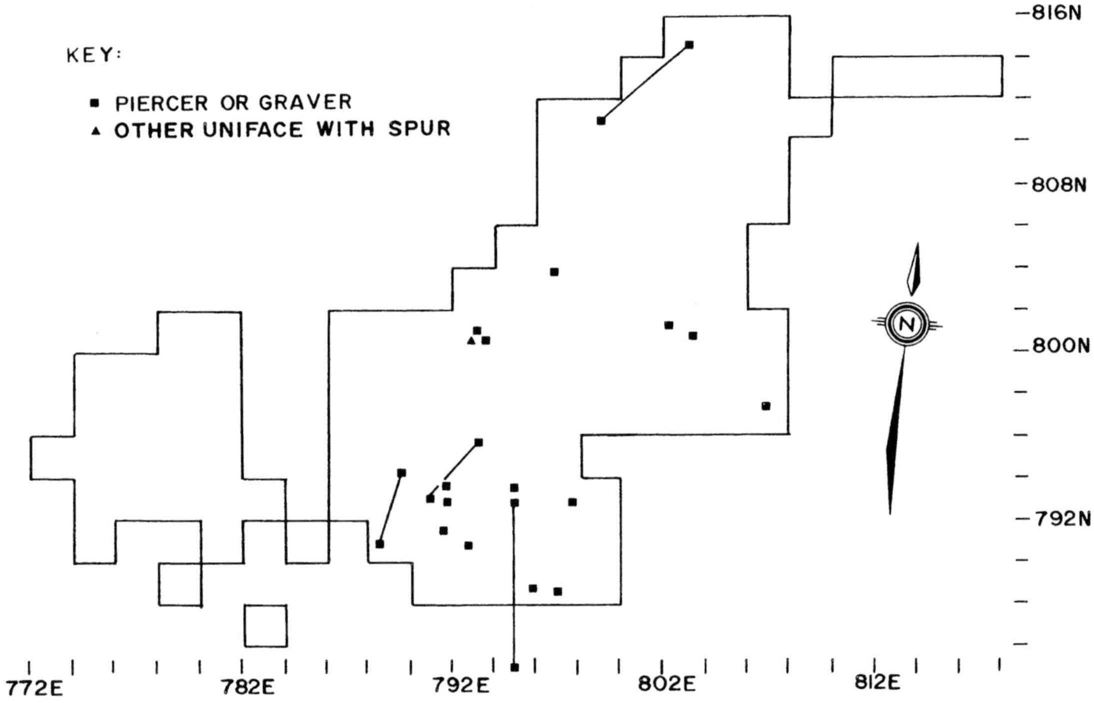

Figure 87. Distribution of piercers, Thedford II site.

Chapter 8

Thedford II in the Fluted Point Tradition

In this chapter, we briefly examine possible relationships of the Thedford II site to other Early Paleo-Indian manifestations east of the Rocky Mountains. The focus is on tool form comparisons (particularly points) and temporal/spatial relationships.

In order to aid discussions, a time/space chart showing suggested relationships of point styles/complexes across North America is provided (Fig. 88). We begin discussion with a consideration of the place of the Thedford II site in the Parkhill complex of the eastern Great Lakes, and gradually expand comparisons to a more comprehensive geographical area. Much of what follows is speculative since factors such as activity variation between sites cannot be easily accounted for with present evidence. Nonetheless, we believe the discussion will be of value, if only to suggest problems, ideas or impressions which can be tested or used as a guide for future research.

The Thedford II Site and the Parkhill Complex

As discussed earlier, at least three fluted point types can be, without any doubt, demonstrated to occur in the eastern Great Lakes area: Gainey, Barnes and Crowfield. We consider these to represent a time series. Indeed, as our analyses have shown, most of the variables which serve to distinguish each type follow patterns consistent with an interpretation of temporal variation.

Since the Thedford II points are Barnes points, the site is assigned to the Parkhill complex or industry, first delineated by Roosa (1977a, 1977b). However, our analyses of Barnes points from several locations indicate that the Parkhill complex points are not as homogeneous as previously thought. In particular, there are significant size differences between the points from Thedford II/Parkhill/Barnes and those from the Fisher site. On this basis we suggested the Fisher site was later in time. Notable in this regard are the distinctive differences in basal and maximum widths and, to a certain extent, basal concavity depth and banding orientation between Fisher and the other sites. Indeed, these may be different enough to warrant the assignment of a different type name to the Fisher site materials and finds with apparently comparable material such as: the McLeod site (Ellis 1984), the Mawson site (Fig. 2, #7; see Deller 1976b: Plate 5m), the Arkona/Stott Glen site (Fig. 2, #8; Deller 1988), and the Grand Traverse County, Michigan, find (Dekin 1966). Accepting some time depth to Barnes points, the larger points from Thedford II lead us to suggest it is an early representative of the Parkhill complex.

The age of the Parkhill complex is a matter of debate but it probably falls around the middle of the occupation of the area by fluted point-producing groups; that is, within one or two hundred radiocarbon years of 10,600 B.P. This complex occurs throughout southwestern Ontario to south-central Ontario, eastern Michigan, northern Ohio (Payne 1982) and upstate New York (e.g., Lothrop 1988:263; Ritchie 1957: Plates 10Ab). Its distribution beyond this area is unknown. Yet, as will be discussed later, points similar to Barnes occur over a much wider area.

The Thedford II investigations have documented the association of several tool types and classes with the Parkhill industry. Tools such as piercers (including coronet forms), trianguloid end scrapers and the general side scraper class are forms known from Paleo-Indian sites over a wide area. Hence, they are not particularly diagnostic or distinctive. However, continued detailed analysis of more assemblages might yield information on distinctive regional or temporal aspects of these tool forms. For example, there are indications of a consistent association of large spurs or borers with side scrapers having concave edges on sites in the area. This consistent association does not appear to occur farther afield (e.g. MacDonald 1968: Plate XVII: b, c).

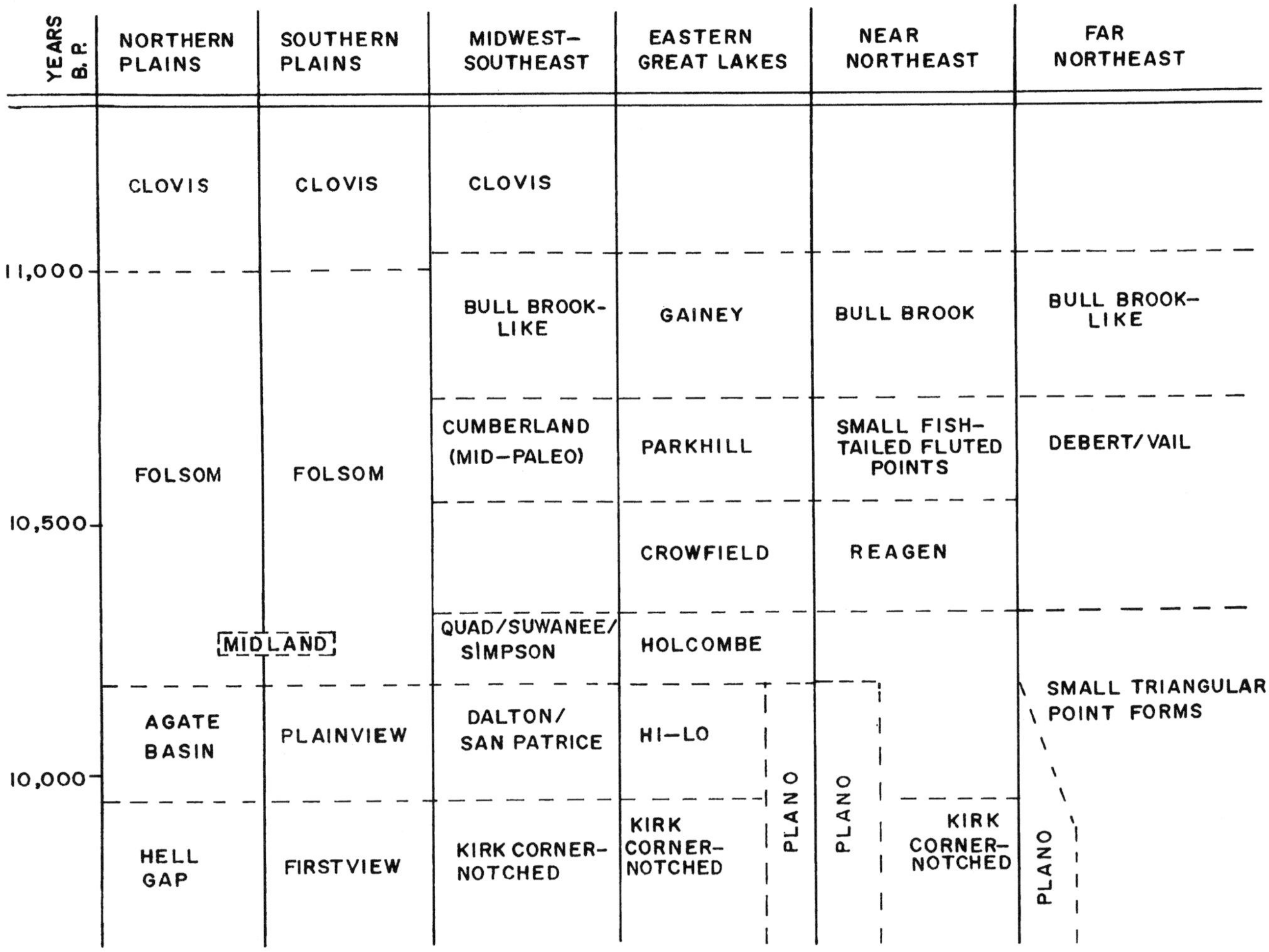

Figure 88. Suggested temporal/spatial relationships of Paleo-Indian/Early Archaic material.

Other tool forms associated with Barnes points at the Thedford II site are apparently much more restricted in distribution. Channel flake points, large alternately beveled bifaces, backed bifaces, narrow end scrapers, offset end scrapers, backed and snapped unifaces and combination notch/borers/denticulates, are tool forms which are known from other Parkhill industry sites. Forms such as proximal end and side scrapers and combination narrow/wide end scrapers are definitely known only from Thedford II among the relatively well-known or excavated sites of the Parkhill complex. However, two apparent proximal end and side scrapers may have been recovered from the Banting site (Storck 1979: Plate 17j, 26h). These could be associated with the Parkhill complex component on that site.

The Parkhill Complex in the Eastern Great Lakes Area

As indicated earlier, the Parkhill complex appears to fall in time between the Gainey and Crowfield complexes of the eastern Great Lakes. Some comparisons with these other complexes are drawn here.

The spatial extent of the Gainey complex is unknown. Certainly, materials assignable to Gainey occur in much the same areas as the Parkhill complex, including upstate New York (e.g. Lothrop 1988; Ritchie 1957: Plate 10Aa). The points associated with this complex were originally assigned by Roosa (1965) to a Great Lakes Bull Brook type indicating their similarity to the points from the Bull Brook site. Indeed, they are quite similar as will be discussed below. This suggests the

Gainey complex, or more properly, the Bull Brook complex, extends much beyond the Great Lakes.

The tools associated with Gainey points are less known than those associated with the Parkhill complex. Excluding widespread tool forms common to many industries, it is possible that both share more specific forms such as offset end scrapers (Simons et al. 1984: Fig. 4e) and tools resembling "Mungai knives" (Simons et al. 1984: Fig. 9b). Here, however, known commonalities end. Tools such as proximal end and side scrapers, narrow end scrapers, backed and snapped unifaces, backed bifaces, alternately beveled bifaces and channel flake points reported from Parkhill contexts, have not been reported in association with Gainey points. Of course, as more Gainey sites are discovered and analyzed, these tool forms may prove to be associated with Gainey points.

We suspect that two forms not common or absent on Parkhill complex sites, may be more common on Gainey sites (see below). One such form is the burin which is reported from the Gainey site (Simons et al. 1984: Fig. 9d) as well as sites we would classify as Gainey complex in northern Ohio such as Nobles Pond (Gramly and Summers 1986). A second form is the *pièce esquillée*. We hesitate to refer to these as tools since there is some evidence they are cores (Goodyear 1982b). Regardless, these artifacts are rare to nonexistent on Parkhill industry sites. One *possible* example occurs in the Thedford II collections. However, this identification is uncertain. In any case, the Thedford II item is not of the classic form reported from sites like Debert or Vail. Other than this possible example, *pièces esquillées* are unreported from Parkhill complex sites such as Parkhill, Barnes, Dixon, McLeod, etc. They can not be considered a classic or common artifact form on sites of this complex.

Although no *pièces esquillées* are reported from the Gainey site, they occur at sites with Gainey components in the Great Lakes area such as Nobles Pond (Gramly and Summers 1986) and Potts (Lothrop 1988:357–58). In Ontario, a Gainey point reworked as a *pièce esquillée* is known to us from a local collection (Fig. 32, #14). Moreover, good examples of this type were recovered from the Ferguson and Weed sites (Ellis and Deller 1980: Fig. 10a; Fig. 32, #'s. 6, 13, this report). These sites have each yielded Gainey points although a single Barnes point also occurs at Weed. Two other examples of *pièces esquillées* occur in the small assemblage from the Murphy site (Deller 1979: Location 36) in southwestern Ontario. We suspect this is a Gainey site but no points have been recovered which could confirm this inference.

The Crowfield complex also occurs in areas occupied by the Parkhill complex, including both western New York (Ritchie 1957: Plate 10Ac) and western Pennsylvania (Lantz 1984: Fig. 7d). We suspect it may extend farther to the east. Elsewhere (Deller and Ellis 1984:50), we have noted the similarity of Crowfield points to the fluted points recovered from the Reagen site in Vermont.

The large number of artifacts from a presumed grave cache at the Crowfield site documents the association of Crowfield points with distinctive tool forms also found in the Parkhill complex. These include: large alternately beveled bifaces, backed bifaces, narrow end scrapers, beaks, and concave side scrapers with spurs. Other forms such as channel flake points, offset end scrapers and proximal end and side scrapers are not known from Crowfield contexts.

Conversely, some forms found in the Crowfield complex are not known from the Parkhill complex. These consist of shouldered fluted points (probably "knives"), leaf-shaped bifaces and rodlike bifaces which may be drills.

A single example of this last form may have been recovered from the Parkhill site ("fluted drill"; Roosa 1977b: Fig. 4). However, there are several problems with assigning this tool to the Parkhill industry. It was recovered much removed from any of the main, well-defined and delimited concentrations at Parkhill. It may be a fluted point which was reworked by later site occupants. Indeed, it closely resembles Early Woodland Meadowood drills recovered from the site. There is also some question as to whether or not the tool is on the diagnostic Collingwood chert and thus, of Paleo-Indian derivation. It is important to note as well that this tool may not be a drill. The tip is somewhat symmetrical and pointed. As Roosa (1980:16) notes, this tool resembles the heavily reworked points reported from the Dalton complex which are variously referred to as "perforators," "skewers," or "scrapers," and are not believed to have been used as drills (see Price and Krakker 1975; Morse 1971:10, 1975:138, 1976:162; Goodyear 1974:30). In any case, even if this tool is associated with the Parkhill complex, we do not know if the incomplete rodlike bifaces at Crowfield are functional equivalents or if they were fluted.

Great Lakes Complexes in Relation to the Northeast

Most of the major sites throughout the near Northeast (i.e., Vermont, southern New England, and easternmost Pennsylvania/New York) appear to be quite similar. Grimes et al. (1984) have suggested that sites such as Bull Brook, Bull Brook II, Whipple and Wapanucket #8 can be grouped together as a "Bull Brook phase." The points from these sites are relatively homogeneous. They are: (1) somewhat large in terms of maximum and basal width; (2) can have deep basal concavities relative to Barnes points; (3) lack or only have slightly developed "fishtails"; (4) are generally parallel-sided with maximum width between midsection and base; (5) appear to be generally well-fluted although fluting is often partial; (6) tend to have single flutes; (7) exhibit a short, parallel to slightly expanding retouch around the basal concavity; (8) frequently exhibit the "Barnes basal finishing technique" (Roosa 1965:96; see especially, Grimes 1979: Fig. 9–1, 9–3); and (9), are often resharpened to quite short "sub-triangular" forms (i.e. Curran 1984: Plate 1–1; Grimes 1979: Fig. 8). In all of these formal characteristics, including the resharpening to sub-triangular forms (see Fig. 27b; Roosa 1965: Plate 1d; Mason 1958: Plate 1b), they are comparable to the Gainey points of the Great Lakes region. By extension, they probably date to a similar time.

Some C-14 dates are available for Bull Brook phase components. The dates obtained from the Bull Brook site of ca. 9000 B.P. (Byers 1959) are much too recent. Accelerator dates from the Whipple site have been reported (Haynes et al. 1984: Table 2). These results suggest two separate charcoal populations were dated. Haynes et al. (1984: 185) suggest the earliest two dates (averaging 11,050 ± 300) are probably a more realistic measure of the site's age.

Extensive sites in the "near Northeast" with points comparable to those of the Parkhill complex of the Great Lakes have yet to be reported. Certainly, finds of relatively small, fishtailed points with relatively shallow basal concavities and lateral basal edges which expand from the base or "waist" are common in this area (e.g. Kraft 1973: Plate 1b-c, 1f; Grimes et al. 1984: Plate 6–1; Kopper et al. 1980: Plate 4–10; and perhaps, Moeller 1980: Plate 7) but large extensive sites are not reported.

As for materials comparable to Crowfield points, sites are present with similar point forms such as Reagen (Ritchie 1953) and Plenge (Kraft 1973: Plate 3). Moreover, we have personally examined several unreported examples from locations in eastern Pennsylvania.

Moving farther afield, materials comparable to Gainey–Bull Brook points do occur in the far Northeast (i.e., Maine and the Maritimes). This is based largely on the occasional point find (e.g. Spiess and Wilson 1977: Plate A3–4; Turnbull 1974) and one or two sites (Bonnichsen et al. 1983; Payne 1987). Yet, it is worth stressing that all these Bull Brook–like finds are from Maine and western New Brunswick with no examples reported farther east into Nova Scotia/Prince Edward Island. These finds exhibit several similarities to the Bull Brook site points including: wide maximum and basal dimensions, roughly parallel sides, a tendency for single fluting, little development of fishtails, a fine retouch in the basal concavity, and so on. On the other hand, many of these characteristics are also to be found among the better known Vail/Debert site assemblages (see MacDonald 1968; Gramly 1982). The exceptions are a tendency for shorter flutes and of course, the very deep basal concavities which average around 9 mm at Vail (Gramly 1982: Table 2). Whatever the case, on present data the Vail/Debert forms do not seem to have a clear analog in the Great Lakes area.

We suspect the Vail/Debert site materials are later in time than Bull Brook materials because they are formally similar to, and appear to lead into, a series of Late Paleo-Indian, basally thinned, trianguloid points (Keenlyside 1985). These trianguloid points apparently become dominant in east coastal areas from Labrador (McGhee and Tuck 1975) to at least New Jersey (Cavallo 1981) up until about 8000 B.P. Moreover, given the validity of the earlier dates from the Whipple site noted above, the 10,400–10,600 B.P. The C-14 dates (Haynes 1987; MacDonald 1968) from Vail/Debert suggest a more recent age placement. Finally, we believe it is significant that fluted points other than those closely conforming to the Vail/Debert forms have never been reported from eastern New Brunswick to Prince Edward Island. Indeed, the lack of significant variation in points in that area is remarkable when compared to the variability evident just to the west and south. It can be suggested that this indicates this area was the last in the Northeast to be colonized by fluted point users. In short, the area was not occupied for long before fluting was abandoned as a basal treatment. Hence, there was little time depth to the fluted point occupation and consequently, little time within which change in fluted point forms could occur.

These age estimates for Vail/Debert materials suggest they are roughly contemporary with the Parkhill and Crowfield complexes of the eastern Great Lakes. In fact, and although the odd Barnes-like point is reported from southernmost Maine (e.g. Spiess and Wilson 1987: Fig. A3–3, Plate A3–6), we have suggested elsewhere (Deller and Ellis 1988) that the lack of clear Parkhill/Crowfield analogs over much of the far Northeast and, conversely, of Vail/Debert points to the west, is a product of this contemporaneity.

In general then, there are suggestions of fluted point

forms reminiscent of the Great Lakes materials throughout parts of the Northeast but several contrasts seem evident the more removed the area is east of the Great Lakes. As for the rest of the tool kit, excluding widespread general forms, there are specific tool types which occur on both Great Lakes and Northeastern sites. For example, the channel flake points found in the Parkhill complex are reported with Bull Brook (Grimes 1979: Fig. 9–5) and probably, Debert/Vail (MacDonald 1968: Plate 6g) materials. Overall, however, we are much more impressed with the differences between Northeastern material and Great Lakes material. We recognize of course, the possibility of biases in the available samples due to activity or "functional" variation. Yet, the fact several sites are now known from both areas tends to confirm variations in the tool kits between the two areas unrelated to biases in activity representation. These may be more indicative of regional variation of a "social" or overall "functional" nature.

First, we are highly impressed with the seeming homogeneity of classes of tools such as end scrapers and side scrapers on many Northeastern sites. For example, at sites like Debert, Vail, Whipple, Bull Brook and Bull Brook II, end scrapers generally seem to be solely of the "trianguloid" type with various degrees of notch/spur development. Consistent varieties seen in the Thedford II and other Great Lakes assemblages seem to be rare or nonexistent. This absence however, might be related to other factors. Most of these Northeastern sites are large in terms of frequencies of artifacts. Consequently, for practical reasons, there has been a tendency to "generalize" assemblage descriptions. Also, even though the Great Lakes types such as narrow end scrapers and proximal end and side scrapers may account in total for a relatively large part of the end scraper tool class, individual types are represented by a limited number of specimens. Hence, they are easy to overlook unless assemblages from a number of sites reported in detail are known. On the other hand, having had the opportunity to examine the assemblages from these Northeastern sites, it is our impression that the Great Lakes types do not occur.

Another possible contrast concerns tools on nonsiliceous materials. Tools on these materials such as scraper planes and choppers occur frequently at Debert (MacDonald 1968: 104–7) and Vail (Gramly 1982). These tools may represent distinct regional and temporal variation as they are much less common on sites in the Great Lakes area. It should be emphasized that the rarity of these tools on Great Lakes sites is not due to them being relegated to later occupations which occur on some of the sites. They are simply rare.

Next, there are several artifact types which consistently occur on Northeastern sites but which appear to be almost nonexistent in some Great Lakes complexes. *Pièces esquillées* are one of these forms. As previous discussions have indicated, these are extremely rare at Parkhill complex sites. In contrast, almost every major site throughout the Northeast yields large numbers of these items. These include: Debert (MacDonald 1968:85–88), Vail (Gramly and Lothrop 1982), Bull Brook (Grimes 1979:111), Bull Brook II (Grimes et al. 1984:164), Whipple (Curran 1984:31) and Shoop (Cox 1986:125). We have suggested however, albeit based on minimal data, that this form may be more common on at least Gainey sites in the Great Lakes area compared to later complexes. This inference may apply to burins as well. Burins are not reported from the Parkhill complex but they occur at Gainey and Nobles Pond and are said to occur at the Bull Brook site (Hadleigh-West 1983:84). Granted this inference, it is possible that these items were part of the artifact inventory of the initial Early Paleo-Indian occupation throughout a larger area. Subsequently, these items could have dropped out of use or become a less important component of the inventory in areas such as the eastern Great Lakes.

Another tool form that is nonexistent in the eastern Great Lakes area is the fluted, blunt-bitted, twist drill. Yet, these tools consistently occur at Whipple (Curran 1984: Plate 5), Vail (Gramly 1982:30–31), Debert (MacDonald 1968:81–85), Bull Brook (Grimes 1979: Fig. 3) and other Northeastern sites. In fact, twist drills *per se* are very rare in the Great Lakes area and reported examples consist solely of the unfluted, stemmed forms reported from Banting (Storck 1979: Plate 10). Similarly, limaces or flake-shavers (Grimes and Grimes 1985) are not known to occur on Great Lakes sites but they frequently occur on sites further to the east such as Debert (MacDonald 1968: Plate X), Vail (Gramly 1982:37–40), Bull Brook (Grimes and Grimes 1985), and Whipple (Curran 1984: Plate 5c).

A final contrast concerns tool forms which consistently occur (albeit in low quantities) on Great Lakes sites but which appear to be absent in the Northeast. These include not only some distinctive end scraper types alluded to earlier but also, tool forms such as large alternately beveled bifaces, backed bifaces and backed and snapped unifaces. Backed bifaces, however, may occur at Shoop (Cox 1986:112) and Plenge (Kraft 1973: Plate 7). Also, some of the items referred to as Enterline side scrapers at Shoop (Witthoft 1952: Plate 3-28), appear to be good examples of proximal end and side scrapers.

In summary, it is possible to suggest that much of the

Northeast and the Great Lakes was initially occupied by Early Paleo-Indian groups employing point forms like those from Gainey and Bull Brook. Their artifact inventories included forms like *pièces esquillées* and perhaps, burins. Presumably this occurred around 11,000 B.P. Subsequently, regional manifestations developed. One was the Parkhill complex of the eastern Great Lakes which, based on a number of surface finds and small sites yielding comparable points, probably extended into southern New England. Another is represented by Debert/Vail in the far Northeast (see Ritchie and Funk 1984:3, for a similar viewpoint). This regionalization may have involved not only changes in the formal and technological characteristics of points but also changes in the tool kit. The latter could have included the termination or reduction of use contexts of *pièces esquillées* and other forms in areas such as the eastern Great Lakes. As well, it is possible that new tool forms were incorporated into the tool kit in some areas. Possible examples are the alternately beveled bifaces and narrow end scrapers found on sites in the Great Lakes region.

It is interesting to note that the presumed later, distinct, regional variants such as Parkhill and Debert/Vail are consistently represented by large sites at the most northerly part of the east known to have been occupied by fluted point users. In more southerly areas such as southern New England, large sites of a presumed post–Bull Brook age are nonexistent or as yet, unreported. In fact, sites with point forms in that area comparable to the Barnes points of the eastern Great Lakes are consistently small (e.g. Dutchess Quarry Cave) or if larger, yield a wider diversity of point forms suggestive of long-term occupation. It has long been recognized that large, non-quarry related sites comparable to Bull Brook, Parkhill, Gainey and Debert in the Northeast do not occur in areas to the south. Recently, Meltzer (1984:348–58, 1988) has suggested this contrast is due to "adaptive" variability among Paleo-Indian groups in the east.

The southern areas are seen to have had a mosaic of microenvironments. They could be characterized as species-rich even though each species was represented by few individuals. In contrast, the northern areas were characterized by fewer species even though the population of single species such as caribou may occur in larger numbers. Hence, Meltzer (1984) argues that differences in fluted point site size and occurrence between the two broad areas is a direct product of responses by Paleo-Indians to these two kinds of environments. The smaller, dispersed food supply in the south favored smaller, dispersed groups. Consequently, the archaeological record consists of small, less intensely occupied sites. A more specialized procurement strategy is seen to have developed in the less diverse habitats of the North. This strategy includes larger, single component sites where larger groups consistently camped to exploit a resource such as caribou.

If Meltzer's (1984) ideas have validity or if simply, these differences are indicative of adaptive variation (regardless of specifics), it is possible to suggest that the "southern site pattern" began to occur or spread to at least the more southern parts of the Northeast on a post–Bull Brook time level. This would account for the present lack of large suspected post–Bull Brook sites in southern New England. Elsewhere, we have suggested that this change is associated with the decline of spruce-dominated vegetation covers and a greater degree of forest closure across the Northeast after 10,600 B.P. (Deller and Ellis 1988).

The Eastern Great Lakes and the Midwest/Southeast

For ease of comparison, we lump together the Midwest (i.e. the Mississippi River valley and the southwestern Great Lakes) and the southeastern United States. As indicated in previous discussions, most of this area has produced only small sites and numerous isolated finds (excepting larger quarry-related sites) although recently, some larger sites have begun to be reported from more northerly regions such as Illinois (e.g. Koldehoff 1983). Nonetheless, the general lack of extensive reported assemblages dictates that we confine ourselves largely to a brief discussion of point forms.

Within this broad area, it is not yet possible to recognize clear-cut fluted point industries comparable to those in the eastern Great Lakes. Material referred to as Clovis (cf. Gardner 1983) is reported throughout this area and as far north as Virginia. However, whether or not these points are exact equivalents of Western Clovis points will continue to be debated for some time, particularly since it is becoming clear that Western Clovis is not as homogeneous as previously thought (see below).

At least some of this material, such as that from the Kimmswick site in eastern Missouri (Graham et al. 1981) is highly similar, if not identical, to Western Clovis material and probably should date to a comparable time. Yet, we concur with Roosa (1965) that at least some of the material referred to as "Clovis" is much too well fluted to be properly called by this term. As well, some of these points exhibit the Barnes basal finishing technique which is absent on Clovis in the West. Indeed, in these regards, these points (see, for example, Morse et al. 1964: #'s 6, 40, etc.; Webb 1950: Fig. 11, 1951: Figs. 8, 9, 10a, 11 *left*) appear more similar to the Gainey–Bull Brook materials from farther to the north and east. Following Roosa (1965:98), we would avoid referring to all these points as Clovis and suggest that many are of a post-Clovis age.

Another variety of fluted point from the area is the Cumberland point (Lewis 1954:7–8). Meltzer's (1984: Fig. 28) recent studies have shown that this point type is common in the middle to upper Mississippi drainage and becomes less common to the north, south and east. It is possible that a time-equivalent in these latter areas are the apparently small fluted points of Gardner's (1983; Gardner and Verrey 1979) "Mid-Paleo" division or smaller forms from the lower Mississippi River valley referred to by Morse and Morse (1983: Fig. 37) as "Sedgewick" points. Cumberland points themselves, with their fishtails and extremely well-fluted surfaces may be a temporal equivalent of Barnes points in the eastern Great Lakes. They are similar enough that they might be confused with each other as noted by Meltzer (1984:335) and, in fact, some would consider Barnes points to be simply Cumberland points (e.g. Justice 1987:25).

To us, Cumberland points appear to differ from Barnes points in that they: (1) are larger on average in terms of overall size and thickness (see Roosa 1977b:119); (2) exhibit much larger, generally more flaring ears; (3) have a much more constricted "waist" and thus, expand much more from the base towards midpoint; and (4) apparently have shallower basal concavities. Moreover, Cumberland points lack the Barnes basal finishing technique (Roosa 1965:98) and often, appear to have a fine basal finishing retouch on both faces similar to Folsom (e.g. Morse et al. 1964: nos. 11, 12, 13, 19, 23, 25, 28, 30, 35, 42). A major problem here, however, is that there is no Cumberland point "type" site and hence, no real *detailed* data on point samples of the same which would allow more than simple impressionistic comparisons. More than these impressionistic assessments will be necessary to resolve such typological debates. Be that as it may, until someone provides a detailed body of data on Cumberland point samples from specific Midwestern sites in which the range of variability in such materials is clearly spelled out, we will continue to call the Great Lakes examples Barnes points.

As noted above, beyond points the rest of the tool kits in the Southeast/Midwest are poorly known. We would like to stress, however, that at least some fluted point manifestations in those areas definitely include an extensive true blade industry (e.g. Sanders 1983, 1988; R. Michael Gramly: pers. comm., 1989). At present definitive evidence of these industries cannot be documented with any certainty in the eastern Great Lakes/Northeast, true blades being so rare in reported assemblages as to be essentially fortuitous (Cox 1986). Since a blade industry is also characteristic of the earliest C-14 dated fluted point development (e.g. Western Clovis; see below), it is plausible to suggest some fluted point industries in the Southeast/Midwest predate the Great Lakes materials (see also Meltzer 1984).

The Eastern Great Lakes in Relation to Western Fluted Point Complexes

Traditionally, Western fluted point materials are divided into two time-sequential complexes: Clovis and Folsom. We have suggested the presence of three or more time/space units in various areas of the east. This contrast with the West should not be taken to imply a greater amount of time variability in the east. The Clovis and Folsom complexes were initially defined on much different criteria than, at least, the Great Lakes materials. The distinction among the Western complexes was based partially on nontechnological criteria such as the hunting of different types of game (i.e. mammoth versus bison).

As well, the definitions of Clovis and Folsom points are based on limited and general criteria such as: partial versus full fluting, overall size, and the nature of edge retouch (i.e. gross versus fine). It is becoming increasingly clear however, that technically and morphologically, materials assigned to Clovis or Folsom vary considerably in terms of characteristics other than those traditionally used. For example, variation in Clovis materials has been documented by Bonnichsen (1977) and Frison (1978: Fig. 3.4). While the tendency has been to view this variability as representing contemporaneous spatial differences, in our estimation there is nothing to indicate this is necessarily the case, and time variation is just as likely an explanation for the perceived differences. If this is the case, one might also expect variation in the associated tool kits through time. Therefore, in the comparisons below, we recognize that what we are considering typical of Clovis or Folsom might be restricted only to certain time periods within each.

In general, although true Folsom points occur in the western Great Lakes (see Roosa 1965:95), no exact equivalents of Clovis or Folsom occur in the eastern Great Lakes. Gainey points, at least in terms of size and outline shape, are comparable to Western Clovis. However, as noted above, they differ in several respects, most notably in that Gainey points have longer fluting than typical Clovis, exhibit deeper basal concavities and have the Barnes basal finishing technique. Similarly, Barnes points with their often fully fluted appearance, tendency for single flute removals, generally smaller sizes, and attempts to recycle short preforms broken in fluting into smaller points (see Tunnell 1975) are comparable to Folsom materials. In fact, they are sometimes identified as Folsom (i.e. Lantz 1984: Fig. 8k). On the other hand, there are several contrasts between Folsom and Barnes points.

For example, Barnes points: have the Barnes basal finishing technique; generally lack fine parallel retouch in the basal concavity; do not exhibit the remnants of the basal fluting "nipple" which is often retained on finished Folsom points; lack the very fine continuous marginal retouch characteristic of Folsom; commonly exhibit fishtails which, in contrast to the rarely reported fishtailed Folsom examples (see Tunnell 1975: Fig. 7–14), are often much more pronounced; are much thicker than Folsom points; appear to expand much more from the base towards midpoint; and rarely, exhibit the extensive, specialized preparation for and fluting on one face prior to preparing the second face (Flenniken 1978).

Another contrast, first noted by Storck (1983:84), is the rare occurrence of "splits" or perverse fractures during fluting on Barnes preforms. This error is very rare to nonexistent at sites like Fisher, Parkhill, Barnes and Thedford II. In contrast, Folsom preforms are often broken in this manner (Tunnell 1975; Wilmsen and Roberts 1978:105–7). The low incidence of this breakage on Barnes preforms may be due to the fact they are much thicker than Folsom points. Nonetheless, on the basis of overall size and extent of fluting, we concur with Roosa (1977b) that Barnes points are closer to Folsom in age than to Clovis.

Crowfield points also appear most similar to Folsom, notably in their extreme thinness and the presence of a fine basal retouch on a relatively shallow concavity. Therefore, as with Barnes points, they may equate with Folsom in age. However, there are contrasts between Folsom and Crowfield points. For example, the latter differ in that they expand markedly from the base, can have pentagonal outlines produced by resharpening and sometimes have a shoulder on one edge.

There are some tool forms which occur in both Western and Great Lakes assemblages. These generally are reported from Folsom rather than Clovis contexts. They include: large beveled bifaces, small points and piercers made on channel flakes and specialized end scrapers resembling those in our "narrow" type.

Large beveled bifaces definitely occur in Folsom assemblages (Frison and Bradley 1980:82) and possibly occur in Clovis assemblages (see Huckell 1979). Moreover, at least some of the Folsom specimens have beveling on alternate edges (Irwin and Wormington 1971:30; Frison and Stanford 1982:55). In addition, both Folsom and Parkhill complex examples can be recycled into bend-break tools (Frison and Bradley 1980:81–82).

Small points made on channel flakes have been reported from Folsom sites such as Adair-Steadman (Tunnell 1975) and Elida (Hester 1962:100). As noted earlier, these definitely occur on Parkhill complex sites. The production of piercers or gravers on channel flakes is another trait reported in both the Parkhill complex (Deller 1980: Plate J8; Storck 1978: Plate 5b) and Folsom (Roosa 1968:149). Specialized narrow end scrapers such as those described in this report also occur in Folsom contexts (Frison and Bradley 1980: Fig. 71c, d) where they are sometimes referred to as "chisel gravers" (Judge 1973: Fig. 15).

Another similarity between some Great Lakes and Western complexes is the absence or rare occurrence of artifact forms common in the Northeast. First, at least the Parkhill and Crowfield complexes yield very few, if any, *pièces esquillées*. These are also rare to nonexistent at Clovis and Folsom sites with only the odd exception (e.g. Roberts 1936:25; Roosa 1968:225). Second, tools similar to those referred to as limaces or flake-shavers in the Northeast have not been reported from Western sites or Great Lakes sites. Some tools referred to as "limaces" have been reported from Folsom (Wilmsen and Roberts 1978: Fig. 94) but these are not the same as the distinctive hafted forms reported from the Northeast (Hadleigh-West 1983:374).

Despite the similarities noted above, there are differences between Great Lakes and Western assemblages. First, in the Great Lakes (and in the Northeast), there is little, if any evidence of a blade component such as that present in Clovis (Haynes 1980:116). Second, both Clovis (Hemmings 1970:124; Frison and Stanford 1982:151) and Folsom (Frison and Bradley 1980:71; Wilmsen and Roberts 1978: Fig. 94) appear to include burins. Excepting Gainey complex sites, these are lacking in the Great Lakes area. Finally, there are several other tool forms which occur on sites representative of at least some Great Lakes complexes which do not appear to occur on Western sites. These include: backed bifaces, offset end scrapers, proximal end and side scrapers, and backed and snapped unifaces.

Conclusions

Our review has suggested several possible contrasts and similarities between fluted point manifestations over a large area. Although these are all worthy of additional investigation and much more adequate testing and evaluation, perhaps not surprisingly, the available data suggest that assemblages from the area intermediate in space between the Northeast and the West (i.e. from the Great Lakes area) also tend to be intermediate in content: Northeastern and Western assemblages appear to be the most dissimilar. Artifact forms such as *pièces esquillées,* flake-shavers and twist drills are common in the Northeast whereas these items are probably nonexistent or very rare in the West. Conversely, Western assemblages include tools such as narrow end scrapers and large beveled bifaces which are not reported from the Northeast.

In contrast, the intermediate eastern Great Lakes area assemblages share traits in common with both areas. For example, at least one of the presumed earlier Great Lakes complexes may commonly include *pièces esquillées*. Moreover, fluted points from the area often exhibit the Barnes basal finishing technique and evidence for blade production is lacking. These are traits which also characterize Northeastern sites but are not reported for Western assemblages. Similarly, some Great Lakes assemblages include tool forms such as large beveled bifaces and narrow end scrapers which are unreported from the Northeast but occur in the West.

CHAPTER 9

Summary and Conclusions

Excavations at the Thedford II site in 1981/82 resulted in the exposure of about 500 square meters of an estimated 700 square meter Early Paleo-Indian occupation area. This area was partially co-extensive with later prehistoric occupations. In addition to spatial and feature data, 159 Paleo-Indian tools/preforms and just over 1,000 waste flakes, as well as a number of tool fragments, were recovered.

The age and environmental contexts of the Thedford II site, as well as other Paleo-Indian sites in the area, are not clearly understood. However, we suggest that fluted point sites in the area date to 11,000–10,400 B.P. based on C-14 dates from other sites in the east and the location of known finds relative to dated Late Pleistocene landforms. Comparisons of point forms suggest the Thedford II site dates to around the middle of this occupation span (ca. 10,600 B.P.). Accepting this estimate, the site was probably in a relatively open spruce-dominated vegetation cover. A more precise environmental context may be provided when the geological/paleoenvironmental studies carried out in conjunction with the nearby Parkhill site investigations are completed.

Almost all of the tools and debris from the site are of Collingwood chert from the Fossil Hill formation of south-central Ontario. This material originates some 175 to 200 km northeast of the site. Unaltered original surfaces of the raw materials present on artifacts indicate that outcrops rather than secondary sources of the raw material were employed. The focus on one material source that is frequently located at a large distance from the site of concern is characteristic of most Great Lakes sites. Interpretation of this pattern is beset by difficulties (see Ellis 1984: 344–68, for an extended discussion). Employing the simple criteria of parsimony and practicality, the suggestion is that the distance between site and lithic source locations is indicative of the distance or part thereof covered by groups on an annual cycle of movement to take advantage of various resources not co-extensive in space or in season of availability (see Roosa 1977a; Storck 1984a for similar views). Almost all known Parkhill complex sites dominated by Collingwood chert include small amounts of Michigan Bayport chert tools and waste. This suggests exchange and other forms of social interaction with groups in Michigan who used that chert source.

Analyses of flake blanks made into tools indicate the Collingwood chert was reduced in a systematic rather than a nonpatterned manner. It is suggested that initial quarry blocks were reduced to unidirectional, "conical" block cores from which large "blade-flakes" were produced. An alternative procedure was to reduce the quarry block to a large biface from which flakes suitable for tools could be derived.

The standardization of reduction sequences is an aspect of Paleo-Indian tool production that is often commented on. It is usually suggested that such standardization allows the production of more useable flakes per lithic unit than does a more haphazard reduction (Knudson 1973:134; MacDonald 1968:66). This viewpoint is partially confirmed by ethnographic studies (Gould 1980:125–26). Hence, it would be a useful strategy under conditions of restricted access to lithic material. However, this explanation may be oversimplified as there is evidence among Paleo-Indian industries that standardized core reduction can occur at the quarry where there are no constraints on lithic source access (see Knudson 1973:92; Tunnell 1975:9–10). Savings in time and effort have been suggested as an alternative explanation (Ellis 1984:436–37) as standardized reduction procedures allow the rapid production of useable blanks. As well, a need for flexibility in the transported tool blanks such that each blank could be potentially made into various *complex* tool type configurations away from quarries may have favored this standardization (Ellis 1984:448–49).

Detailed analysis and comparisons of the fluted biface sample from Thedford II with samples derived from several other sites and locations in the eastern Great Lakes region clearly indicate that the Thedford II site should be assigned to the Parkhill complex or industry, of which Barnes points (Roosa 1965) are the diagnostic indicator. Although our major purpose in carrying out these comparisons was to show the essential similarity of the Thedford II bifaces to Barnes points, the analyses proved useful in other respects. The analyses supported the work and contentions of Roosa (1965) and Wright (1981a) that point types distinguished by the recurrent clustering of certain attribute states and restricted variable ranges, which are not a product of factors such as recycling or resharpening, do exist. Moreover, the differences between the Great Lakes types, particularly in terms of continuous variables, largely follow trends which are consistent with an interpretation of a time series rather than contemporaneous variation.

We believe that these are important insights and would hope that they would encourage work along similar lines in the East. We would suggest, following Roosa (1965), that the largely impressionistic and arbitrary assignment of Eastern forms to broad categories such as Clovis, based on a consideration of a minimal number of characteristics, is misleading or directs attention away from the search for patterning in the complexity of Eastern fluted point forms. Consequently, it directs attempts away from the development of more refined cultural-historical frameworks or the attainment of more precise means to evaluate inter-assemblage relationships.

The analyses of the rest of the Thedford II assemblage resulted in the recognition of several tool types and classes. Several of the types such as alternately beveled bifaces, backed bifaces, offset end scrapers, proximal end and side scrapers, narrow end scrapers, and backed and snapped unifaces have not been widely recognized in the past or have never before been reported. Given their distinctiveness, they should be useful for recognizing Paleo-Indian sites in the absence of traditionally used diagnostics. Indeed, they have already proved useful in this regard in our own fieldwork. Also, these tool forms may be spatially and/or temporally restricted in distribution. Thus, they may be indicative of regional variability in Paleo-Indian tool kits.

The highly distinctive nature of these tool forms also strongly suggests that they were made for use in certain task contexts (Ellis and Deller 1988). Therefore, they should be of more value as measures of activity variation on sites. In addition, they suggest Paleo-Indian tool kits may not always be small and undiversified as some investigators have implied. We would suggest this belief in limited tool kits is partially a product of the use of typologies in which tools are classified into broad descriptive units such as end scrapers, side scrapers and retouched flakes. These units are created as cultural-historical measures or for descriptive purposes, and should not be seen as use categories.

The lithic debris analysis indicated lithic raw materials were transported to the site as finished tools, flake blanks and biface preforms. This is a characteristic of all known sites away from quarries and is best seen as a strategy to reduce the bulk of the lithic material to be transported a great distance. It avoids the transportation of large amounts of material which will be trimmed off and discarded as waste. Moreover, some, if not the majority, of the transported tool kit was maintained in a more "flexible" state of flake blanks or preforms. Thus, each tool blank/preform could serve for or be made into different tool types suitable for different uses. This flexibility would be valuable in situations of unpredictable tool use requirements at times and locations removed from lithic procurement events.

The debris analysis also provided the opportunity to suggest problems with the use of debris-to-tool ratios. These ratios have often played a central role in the interpretation of interassemblage variability, particularly with regard to time elapsed since lithic procurement. However, it was suggested that there are other factors influencing these ratios which must be considered before the place of various sites in relative positions versus time of lithic procurement can be achieved. Besides the problem of different recovery techniques, a major problem with these ratios concerns the degree of emphasis on the reduction of bifacial versus unifacial tools. In particular, if there is a greater emphasis on biface reduction relative to uniface reduction on a site, we should expect greater amounts of debris to be produced.

In addition to this effect, the presence of large amounts of debris as the result of biface reduction: (1) may skew the debris densities at the loci where uniface production and use was important; or (2) may bias excavation to areas of biface production, if one uses surface debris densities to determine where to excavate. It was also suggested that as the frequency of biface tools in an assemblage increases, one might also expect increases in the debris-to-tool ratios. The limited number of assemblages for which data are available support this suggestion. However, exceptions do occur and it was demonstrated that these exceptions can have important interpretive potential. For example, despite the high percentage of bifaces in A-northeast at Thedford II, there was a relatively low amount of debris. This exception provided an independent test and support of the suggestion derived from the biface analysis that a cache of fluted bifaces was present in this area. In sum, the assemblage included a number of tools "discarded" but not reduced at that locus.

Well-defined Paleo-Indian features were not documented at the site. However, based on subsoil anomalies and density of artifactual material, several features are suggested. Due to their age, these features had lost visible outlines and/or had been disturbed greatly by later prehistoric groups, by European agricultural activities, or by rodent-root action.

The spatial analyses of the distribution of items resulted in the recognition six clusters of artifactual material. Five of these, arranged in a semicircle around the site periphery, are similar in size, artifact yield, and feature content. Based on variability in the use of raw materials from different sources and variability in the fluted biface samples from these clusters, it is suggested they were each occupied by different individuals and perhaps, by different small social units such as households. The final, central artifact cluster was larger, yielded more tools, and the suspected features were different. Its size, central position and artifact content suggest the area was a central communal work area. Regardless of whether one accepts these interpretations of spatial patterning, it is important to note that this patterning was still present even though most of the site had been plowed for some time.

Finally, we carried out comparisons to place the Thedford II site into the wider context of the fluted point-associated occupations of most of North America. It is suggested that the Parkhill complex, of which Thedford II is a representative, falls around the middle of the Early Paleo-Indian occupation of the area. Moreover, it is inferred that this complex is a rough contemporary of Folsom in the West; Cumberland in the upper Mississippi drainage; the "Mid-Paleo" of the mid-Atlantic to southeastern United States; a series of small, well-fluted, fishtailed points in southern New England; and the Debert/Vail material of northern Maine and the Canadian Maritimes. There are suggestions that the tool kits associated with these and other fluted point forms may vary throughout North America. Much more research needs to be accomplished in order to test this proposition.

Bibliography

Adovasio, J., J. Donahue, K. Cushman, R. Carlisle, R. Stuckenrath, J. Gunn and W. Johnson
1983 Evidence from Meadowcroft Rockshelter. *In* Early Man in the New World, R. Shutler Jr., ed., pp. 163–89. Beverly Hills: Sage Publications.

Ashworth, A.C.
1977 A Late Wisconsin *Coleopterous* assemblage from southern Ontario and its environmental significance. Canadian Journal of Earth Sciences 14:1625–34.

Binford, L.R.
1972 Archaeological systematics and the study of culture process. *In* An Archaeological Perspective, by L.R. Binford, pp. 195–207. New York: Academic Press.

Binford, L.R., and G.I. Quimby
1972 Indian sites and chipped stone materials in the western Lake Michigan area. *In* An Archaeological Perspective, by L.R. Binford, pp. 346–72. New York: Academic Press.

Bonnichsen, R.
1977 Models for Deriving Cultural Information From Stone Tools. National Museum of Man, Archaeological Survey of Canada, Mercury Series, Paper No. 60.

Bonnichsen, R., B.J. Borque and D.E. Young
1983 The Moosehorn fluted point discovery, northern Maine. Archaeology of Eastern North America 11:36–48.

Bordes, F.
1961 Typologie du Paleolithique, Ancien et Moyen. Publications de L'Institut de L'Universite de Bordeaux, Memoire No. 1.
1972 A Tale of Two Caves. New York: Harper & Row.

Bordes, F., and D. Crabtree
1969 The Corbiac blade technique and other experiments. Tebiwa 12(2):1–13.

Bradley, B.
1974 Comments on the lithic technology of the Casper site. *In* The Casper Site: A Hell Gap Bison Kill on the High Plains, G. Frison, ed., pp. 191–97. New York: Academic Press.
1982 Flaked stone technology and typology. *In* The Agate Basin Site: A Record of the Paleo-Indian Occupation of the Northwestern High Plains, G. Frison and D. Stanford, eds., pp. 181–208. New York: Academic Press.

Brink, J.
1978 An Experimental Study of Microwear Formation on End Scrapers. National Museum of Man, Archaeological Survey of Canada, Mercury Series, Paper No. 83.

Brown, J., and C. Cleland
1968 The late glacial and early postglacial faunal resources in biomes newly opened to human adaptation. *In* The Quaternary of Illinois, R. Bergstrom, ed., University of Illinois College of Agriculture, Special Publication No. 14:114–22.

Byers, D.
1959 Radiocarbon dates for the Bull Brook site, Massachusetts. American Antiquity 24(4):427–29.

Callahan, E.
1979 The basics of biface knapping in the Eastern fluted point tradition: A manual for flint-knappers and lithic analysts. Archaeology of Eastern North America 7(1):1–180.

Cavallo, J.
1981 Turkey Swamp: A Late Paleo-Indian site in New Jersey's coastal plain. Archaeology of Eastern North America 9:1–18.

Clarke, D.L.
1968 Analytical Archaeology. London: Methuen and Co.

Cleland, C.
1965 Barren ground caribou (*Rangifer arcticus*) from an Early Man site in southeastern Michigan. American Antiquity 30(3):350–51.
1966 The Prehistoric Animal Ecology and Ethnozoology of the Upper Great Lakes Region. Anthropological Papers, Museum of Anthropology, University of Michigan No. 29.
1982 The inland shore fishery in the Upper Great Lakes: its development and importance in prehistory. American Antiquity 47(4):761–84.

Collins, M.B.
1975 Lithic technology as a means of processual inference. *In* Lithic Technology: Making and Using Stone Tools, E. Swanson, ed., pp. 15–34. The Hague: Mouton.

Cooper, A.J.
1979 Quaternary Geology of the Grand Bend—Parkhill Area, Southern Ontario. Ontario Geological Survey, Report No. 188.

Cotterell, B., and J. Kamminga
1979 The mechanics of flaking. *In* Lithic Use-Wear Analysis, B. Hayden, ed., pp. 97–112. New York: Academic Press.

Cox, S.L.
1986 A re-analysis of the Shoop site. Archaeology of Eastern North America 14:101–70.

Curran, M.
1984 The Whipple site and Paleo-Indian tool assemblage variation: a comparison of intrasite structuring. Archaeology of Eastern North America 12:5–40.

Dekin, A.A.
1966 A fluted point from Grand Traverse county. Michigan Archaeologist 12(1):35–36.

Deller, D.B.
1976a Paleo-Indian locations on Late Pleistocene shorelines, Middlesex county, Ontario. Ontario Archaeology 26:3–19.
1976b The Heaman site: a preliminary report on a Paleo-Indian site in Middlesex county, Ontario. Ontario Archaeology 27:13–28.
1979 Paleo-Indian reconnaissance in the counties of Lambton and Middlesex, Ontario. Ontario Archaeology 32:3–20.
1980a The Parkhill (Brophey) site, AhHk-49: analyses of surface materials. Unpublished MA thesis, Wilfrid Laurier University, Waterloo, Ontario.
1980b An archaeological survey in the counties of Lambton and Middlesex including test excavations at two Paleo-Indian sites. Report on file, Ministry of Citizenship and Culture, Toronto, Ontario.
1988 The Paleo-Indian occupation of southwestern Ontario: distribution, technology and social organization. Unpublished PhD dissertation, McGill University, Montreal, Quebec.
1989 Interpretation of chert type variation in Paleoindian industries, southwestern Ontario. *In* Eastern Paleoindian Lithic Resource Use, C.J. Ellis and J.C. Lothrop, eds., pp. 191–220. Boulder: Westview Press.

Deller, D.B., and C.J. Ellis
1984 Crowfield: a preliminary report on a probable Paleo-Indian cremation in southwestern Ontario. Archaeology of Eastern North America 12:41–71.
1988 Early Paleo-Indian complexes in southwestern Ontario. *In* Late Pleistocene and Early Holocene Paleoecology and Archaeology of the Eastern Great Lakes, R.S. Laub, N. Miller and D. Steadman, eds. Bulletin of the Buffalo Society of Natural Sciences 33:251–63.

Deller, D.B., C.J. Ellis and I.T. Kenyon
1986 Archaeology of the southeastern Huron Basin. *In* Studies in Southwestern Ontario Archaeology, Occasional Publications of the London Chapter, Ontario Archaeological Society, W.A. Fox, ed., No. 1:3–12.

DeVisscher, J., E. Wahla and J. Fitting
1970 Additional Paleo-Indian campsites adjacent to the Holcombe site. Michigan Archaeologist 16(1):1–23.

Dreimanis, A.
1968 Extinction of mastodons in eastern North America: testing a new climatic-environment hypothesis. Ohio Journal of Science 68(6):257–72.

Dustin, F.
1935 A study of Bayport chert. Papers of the Michigan Academy of Science, Arts and Letters 20:465–75.

Eisenberg, L.
1978 Paleo-Indian Settlement Patterns in the Hudson and Delaware River Drainages. Occasional Papers in Northeastern Anthropology No. 4.

Eley, B., and P. Von Bitter
1985 Reports on "Collingwood" and "Balsam Lake" cherts as they apply to the Deavitt site. Appendix II in Late Paleo-Indian Settlement Patterns Along the Margins of the Simcoe Lowlands in Southcentral Ontario, by G.C. Dibb, pp. 255–59. Unpublished MA thesis, Dept. of Anthropology, Trent University, Peterborough, Ontario.

Ellis, C.J.
1979 Analysis of Lithic Debitage From Fluted Point Sites in Ontario. Unpublished MA thesis, Department of Anthropology, McMaster University, Hamilton, Ontario.
1984 Paleo-Indian Lithic Technological Structure and Organization in the Lower Great Lakes Area: A First Approximation. Unpublished PhD dissertation, Simon Fraser University, Burnaby, BC.

Ellis, C.J., and D.B. Deller
1980 Test excavations on two small fluted point sites in Ontario. *In* Deller (1980b):92–156.
1982 Hi-Lo materials from southwestern Ontario. Ontario Archaeology 38:3–22.
1986 Post-glacial Lake Nipissing "Waterworn" assemblages from the southeastern Huron Basin area. Ontario Archaeology, 45:39–60.
1988 Some distinctive Paleo-Indian tool types from the Lower Great Lakes area. Midcontinental Journal of Archaeology 13:111–58.

Ericson, J.F.
1984 Toward the analysis of lithic production systems. *In* Prehistoric Quarries and Lithic Production, J. Ericson and B. Purdy, eds., pp. 1–19. Cambridge: Cambridge University Press.

Eschman, D.F., and P.F. Karrow
1985 Huron Basin glacial lakes: a review. *In* Quaternary Evolution of the Great Lakes. Geological Association of Canada, Special Paper 30:79–93.

Fitting, J.E.
1963 The Hi-Lo site: a late Paleo-Indian site in Michigan. Wisconsin Archaeologist 44(2):87–96.

Fitting, J.E., J. De Visscher and E. Wahla
1966 The Paleo-Indian Occupation of the Holcombe Beach. Museum of Anthropology, University of Michigan, Anthropological Papers No. 27.

Flenniken, J.J.
1978 Reevaluation of the Lindenmeier Folsom: a replication experiment in lithic technology. American Antiquity 43(3):473–80.

Frison, G.C.
1968 A functional analysis of certain chipped stone tools. American Antiquity 33(2):149–55.
1978 Prehistoric Hunters of the High Plains. New York: Academic Press.

Frison, G., and B. Bradley
1980 Folsom Tools and Technology at the Hanson Site, Wyoming. Albuquerque: University of New Mexico Press.

Frison, G., and D. Stanford, eds.
1982 The Agate Basin Site: A Record of the Paleo-Indian Occupation of the Northwestern High Plains. New York: Academic Press.

Fullerton, D.S.
1980 Preliminary Correlation of Post-Erie Interstadial Events (16,000 to 10,000 Radiocarbon Years Before Present), Central and Eastern Great Lakes Region and Hudson, Champlain and St. Lawrence Lowlands, United States and Canada. United States Geological Survey, Professional Paper 1089.

Funk, R.E.
1973 The West Athens Hill site. *In* Aboriginal Settlement Patterns in the Northeast by W.A. Ritchie and R.E. Funk, pp. 9–36. New York State Museum and Science Service, Memoir No. 20.
1982 Introduction. *In* The Vail Site a Paleo-Indian Encampment in Maine by R.M. Gramly, pp. xii-xiv, Bulletin of the Buffalo Society of Natural Sciences, Volume 30.

Funk, R.E., T. Weinman and P. Weinman
1969 A recently discovered Paleo-Indian manifestation in Greene county, New York. New York State Archaeological Association, Bulletin 45:1–23.

Gardner, W.
1974 The Flint Run complex: pattern and process during the Paleo-Indian to Early Archaic. Archaeology Laboratory, Department of Anthropology, Catholic University of America, Occasional Paper 1:5–47.
1976 Paleo-Indian to Early Archaic: continuity and change in eastern North America during the Late Pleistocene to Early Holocene. IX Union Internationale Des Sciences Prehistoriques et Protohistoriques, pp. 19–46. Nice, France.
1977 Flint Run Paleo-Indian complex and its implications for eastern North American prehistory. Annals of the New York Academy of Sciences 288:157–263.
1983 Stop me if you've heard this one before: the Flint Run complex revisited. Archaeology of Eastern North America 11:49–64.

Gardner, W., and R. Verrey
1979 Typology and chronology of fluted points from the Flint Run area. Pennsylvania Archaeologist 49(1):13–46.

Garrad, C.
1971 Ontario fluted point survey. Ontario Archaeology 16:3–18.

Goodyear, A.C.
1974 The Brand site: A Techno-Functional Study of a Dalton Site in Northeast Arkansas. Arkansas Archaeological Survey, Research Series 7.
1982a The chronological position of the Dalton horizon in the southeastern United States. American Antiquity 47(2):382–95.
1982b Tool Kit Entropy and Bipolar Reduction: A Study of Interassemblage Lithic Variability among Paleo-Indian Sites in the Northeastern United States. Ms. on file, Institute of Archaeology and Anthropology, University of South Carolina, Columbia, S.C.

Gould, R.A.
1980 Living Archaeology. Cambridge: Cambridge University Press.

Graham, R., C.V. Haynes, D.L. Johnson and M. Kay
1981 Kimmswick: a Clovis-mastodon association in eastern Missouri. Science 213:1115–17.

Gramly, R.M.
1982 The Vail Site: A Palaeo-Indian Encampment in Maine. Bulletin of the Buffalo Society of Natural Sciences, Volume 30.

Gramly, R.M., and J. Lothrop
1984 Archaeological investigations of the Potts site, Oswego county, New York, 1982 and 1983. Archaeology of Eastern North America 12:122–58.

Gramly, R., and G. Summers
1986 Nobles Pond: a fluted point site in northeastern Ohio. Mid-Continental Journal of Archaeology 11(1):97–123.

Greco, B.B.
1985 An Investigation of Paleo-Indian Lifeways at the Thedford II Site in Southwestern Ontario. Unpublished MA thesis, University of Manitoba, Winnipeg, Manitoba.

Grimes, J.
1979 A new look at Bull Brook. Anthropology 3:109–30.

Grimes, J., W. Eldridge, B. Grimes, A. Vaccaro, F. Vaccaro, J. Vaccaro, N. Vaccaro and N. Orsini
1984 Bull Brook II. Archaeology of Eastern North America 12:159–83.

Grimes, J., and B. Grimes
1985 Flakeshavers: morphometric, functional and life-cycle analyses of a Paleo-Indian unifacial tool class. Archaeology of Eastern North America 13:35–57.

Hadleigh-West, F.
1983 The antiquity of man in America. *In* Late-Quaternary Environments of the United States, S.C. Porter, ed., Vol. 1::346–2. Minneapolis: University of Minnesota Press.

Hanks, C.C.
1988 The Foxie Otter Site: A Multicomponent Occupation North of Lake Huron. Anthropological Papers, Museum of Anthropology, University of Michigan No. 79.

Hayden, B.
1977 Stone tool functions in the Western desert. *In* Stone Tools as Cultural Markers: Change, Evolution and Complexity, R.V.S. Wright, ed., pp. 178–88. Australian Institute of Aboriginal Studies, Canberra.

Haynes, C.V.
1980 The Clovis culture. Canadian Journal of Anthropology 1(1):115–21.
1987 Clovis origins update. Kiva 52:83–93.

Haynes, C.V., D.J. Donahue, A. Zull and T. Zabel
1984 Application of accelerator dating of fluted point sites. Archaeology of Eastern North America 12:184–91.

Hemmings, E.T.
1970 Early Man in the San Pedro Valley, Arizona. Unpublished PhD dissertation, University of Arizona.

Henry, D.O., C.V. Haynes and B. Bradley
1976 Quantitative variations in flaked stone debitage. Plains Anthropologist 21(1):57–61.

Hester, J.J.
1962 A Folsom complex from the Elida site, Roosevelt county, New Mexico. El Palacio 69:92–113.

Hough, J.
1963 The prehistoric Great Lakes in North America. American Scientist 51:84–109.

Huckell, B.
1979 Of chipped stone tools, elephants and the Clovis hunters: an experiment. Plains Anthropologist 24:177–89.

Irwin, H.T., and H.M. Wormington
1971 Paleo-Indian tool types in the Great Plains. American Antiquity 35(1):24–34.

Jackson, L.
1978 Late Wisconsin Environments and the Palaeo-Indian Occupation in the Northeastern United States and Southern Ontario. Unpublished MA thesis, Department of Anthropology, Trent University, Peterborough, Ontario.
1983 Geochronology and settlement disposition in the Early Paleo-Indian occupation of southern Ontario, Canada. Quaternary Research 19:288–99.

Jelinek, A.J.
1965 Some distinctive flakes and flake tools from the Llano Estacado. Papers of the Michigan Academy of Science, Arts and Letters 51:399–405.

Judge, W.J.
1970 Systems analysis and the Folsom-Midland question. Southwestern Journal of Anthropology 26(1):40–51.
1973 Paleo-Indian Occupation of the Central Rio Grande Valley in New Mexico. Albuquerque: University of New Mexico Press.

Justice, N.D.
1987 Stone Age Spear and Arrow Points of the Midcontinental and Eastern United States. Bloomington: Indiana University Press.

Karrow, P.F.
1980 The Nipissing transgression around southern Lake Huron. Canadian Journal of Earth Sciences 17:1271–74.

Karrow, P.F., T.W. Anderson, A.H. Clarke, L.D. Delorme and M.R. Sreenivasa
1975 Stratigraphy, paleontology and age of Lake Algonquin sediments in southwestern Ontario. Quaternary Research 5:49–87.

Keenlyside, D.L.
1985 Late Paleo-Indian evidence from the southern Gulf of St. Lawrence. Archaeology of Eastern North America 13:79–92.

Kenyon, I.T.
1979 The sub-greywacke lanceolate biface in the Ausable River valley. Kewa, Newsletter of the London Chapter, Ontario Archaeological Society, April, pp. 2–8.
1980 The George Davidson site: an Archaic "broadpoint" component in southwestern Ontario. Archaeology of Eastern North America 8:11–28.

Kidd, K.E.
1951 Fluted points in Ontario. American Antiquity 16(3):260.

Knudson, R.
1973 Organizational Variability in Late Paleo-Indian Assemblages. PhD dissertation, Washington State University, Pullman.

Koldehoff, B.
1983 Paleo-Indian chert utilization and site distribution in southwestern Illinois. Wisconsin Archaeologist 64(3–4):201–38.

Kopper, J., R.E. Funk and L. Dumont
1980 Additional Paleo-Indian and Archaic materials from the Dutchess Quarry Cave area, Orange county, New York. Archaeology of Eastern North America 8:125–37.

Kraft, H.C.
1973 The Plenge site: a Paleo-Indian occupation site in New Jersey. Archaeology of Eastern North America 1(1):56–117.

Lantz, S.
1984 Distribution of Paleo-Indian projectile points and tools from western Pennsylvania: implications for regional differences. Archaeology of Eastern North America 12:210–30.

Larsen, C.E.
1984 Geoarchaeological interpretations of Great Lakes coastal environments. *In* Archaeological Sediments in Context, J. Stein and W. Farrand, eds., pp. 91–110. Center For the Study of Early Man, Edited Volume Series 1.

Laye, G.
1977 A Study of Cherts Used by the Paleolithic Indians in the Parkhill Area with Relationships to the Limestone Containing

Nodular Chert Formations of Southwestern Ontario. Unpublished BSc thesis, Department of Geology, University of Western Ontario, London.

Lewis, C.F.M.
1969 Late quaternary history of lake levels in the Huron and Erie basins. Proceedings of the 12th Conference on Great Lakes Research, pp. 250–70.

Lewis, T.M.N.
1954 The Cumberland point. Oklahoma Anthropological Society, Bulletin 2:7–8.

Lothrop, J.
1986 Activity Structure on Northeastern Paleo-Indian Habitation Sites. Paper presented at the Society for American Archaeology Meetings, New Orleans, Louisiana, April 1986.
1988 The Organization of Paleo-Indian Lithic Technology at the Potts Site. PhD dissertation, State University of New York, Binghamton, New York. University Microfilms, Ann Arbor.
1989 The organization of Paleoindian lithic technology at the Potts site. *In* Eastern Paleoindian Lithic Resource Use, C.J. Ellis and J.C. Lothrop, eds., pp. 99–137. Boulder: Westview Press.

Lothrop, J., and R.M. Gramly
1982 Pieces esquillees from the Vail site. Archaeology of Eastern North America 10:1–22.

Luedtke, B.
1976 Lithic Material Distributions and Interaction Patterns During the Late Woodland Period in Michigan. Unpublished PhD dissertation, University of Michigan.

MacDonald, G.F.
1968 Debert: A Palaeo-Indian Site in Central Nova Scotia. National Museum of Man, Anthropology Papers No. 16.
1971 A review of research on Paleo-Indian in eastern North America, 1960–1970. Arctic Anthropology 8(2):32–41.

Marshall, S.B.
1981 Artifact Form and Function: Implications of Morphological Classification and Wear Pattern Analysis for Cultural Interpretation. Unpublished PhD dissertation, Columbia University, New York.

Mason, R.J.
1958 Late Pleistocene Geochronology and the Paleo-Indian Penetration into the Lower Michigan Peninsula. Anthropological Papers, Museum of Anthropology, University of Michigan No. 11.
1981 Great Lakes Archaeology. New York: Academic Press.

McAndrews, J.
1981 Late Quaternary climate of Ontario: temperature trends from the fossil pollen record. *In* Quaternary Paleoclimate, W.C. Mahaney ed., pp. 319–33. Norwich: Geoabstracts.

McAndrews, J., and J.J. Jackson
1981 Age and environment of Late Pleistocene mastodon and mammoth in southern Ontario. *In* Late Pleistocene and Early Holocene Paleoecology and Archaeology of the Eastern Great Lakes Region, R. Laub, N. Miller and D. Steadman, eds. Bulletin of the Buffalo Society of Natural Sciences 33:161–72.

McCary, B.C.
1951 A workshop site of early man in Dinwiddie county, Virginia. American Antiquity 17(1):9–17.

McGhee, R., and J. Tuck
1975 An Archaic Sequence from the Strait of Belle Isle, Labrador. National Museum of Man, Archaeological Survey of Canada, Mercury Series Paper No. 34.

Meltzer, D.
1984 Late Pleistocene Human Adaptations in Eastern North America. Unpublished PhD dissertation, University of Washington, Seattle, Washington.
1985 On stone procurement and settlement mobility in eastern fluted point groups. North American Archaeologist 6(1):1–24.
1988 Late Pleistocene human adaptations in eastern North America. Journal of World Prehistory 2:1–52.

Moeller, R.W.
1980 6LF21: A Paleo-Indian Site in Western Connecticut. American Indian Archaeological Institute, Occasional Paper No. 2.

Morse, D.F.
1971 The Hawkins cache: a significant Dalton find in northeast Arkansas. The Arkansas Archaeologist 12(1):9–20.
1975 Paleo-Indian in the land of opportunity: preliminary report on excavations at the Sloan site (3GE94). *In* The Cache River Archaeological Project: An Experiment in Contract Archaeology, M. Schiffer and J. House, assemblers, pp. 135–143. Arkansas Archaeological Survey, Research Series No. 8.
1976 An analysis of the Dalton complex in the central Mississippi Valley. Proceedings of the IX Union Internationale Des Sciences Prehistoriques et Protohistoriques, pp. 136–66. Nice, France.

Morse, D.F., and P. Morse
1983 Archaeology of the Central Mississippi Valley. New York: Academic Press.

Morse, D., P. Morse and J. Waggoner Jr.
1964 Fluted points from Smith county, Tennessee. Tennessee Archaeologist 20(1):16–34.

Mott, R.J., and L.D. Farley-Gill
1978 A Late Quaternary pollen profile from Woodstock, Ontario. Canadian Journal of Earth Sciences 15:1101–11.

Muto, G.
1971 A Technologial Analysis of the Early Stages in the Manufacture of Lithic Artifacts. Unpublished MA thesis, Department of Anthropology, Idaho State University.

Newcomer, M.
1971 Some quantitative experiments in handaxe manufacture. World Archaeology 3:84–94.

Odell, G.
1979 A new and improved system for the retrieval of functional information from microscopic observations of chipped stone tools. *In* Lithic Use-Wear Analysis, B. Hayden, ed., pp. 329–44. New York: Academic Press.

Ogden, G.
1977 The Late Quaternary paleoenvironmental record of northeastern North America. Annals of the New York Academy of Sciences 288:26–34.

Ozker, D.
1976 Heat treatment of Bayport chert. Michigan Archaeologist 22(4):357–68.

Payne, J.
1982 The Western Basin Paleo-Indian and Early Archaic Sequences. Unpublished paper.
1987 Windy City (154–16): a Paleoindian lithic workshop in northern Maine. MSc thesis, Institute for Quaternary Studies, University of Maine, Orono.

Price, J., and J. Krakker
1975 Dalton Occupation of the Ozark Border. University of Missouri-Columbia, Museum Briefs No. 20.

Prufer, O., and N. Wright
1970 The Welling site (33CO-2): a fluted point workshop in Coshocton county, Ohio. Ohio Archaeologist 20(4):259–68.

Rice, P.
1981 Evolution of specialized pottery production: a trial model. Current Anthropology 22:219–40.

Ritchie, W.A.
1953 A probable Paleo-Indian site in Vermont. American Antiquity 18(3):249–58.
1957 Traces of early man in the Northeast. New York State Museum and Science Service, Bulletin No. 358.
1980 The Archaeology of New York State. New York: Harbor Hill.

Ritchie, W.A., and R.E. Funk
1984 Paleo-Indians in new perspective: comments on the assembled papers. Archaeology of Eastern North America 12:1–4.

Roberts, A.
1984 Paleo-Indian on the north shore of Lake Ontario. Archaeology of Eastern North America 12:248–65.

Roberts, F.H.H. Jr.
1935 A Folsom Complex: Preliminary Report on Investigations at the Lindenmeier Site in North Colorado. Smithsonian Miscellaneous Collections 94(4).
1936 Additional Information on the Folsom Complex. Smithsonian Miscellaneous Collections 95(10).

Roosa, W.B.
1963 Some Michigan fluted point sites and types. Michigan Archaeologist 9(3):44–48.
1965 Some Great Lakes fluted point types. Michigan Archaeologist 11(3–4):89–102.
1968 Data on Early Sites in Central New Mexico and Michigan. PhD dissertation, University of Michigan. University Microfilms, Ann Arbor.
1977a Great Lakes Paleo-Indian: the Parkhill site, Ontario. Annals of the New York Academy of Sciences 288:349–54.
1977b Fluted points from the Parkhill site. Anthropological Papers, Museum of Anthropology, University of Michigan No. 61:87–122.
1980 The Parkhill complex and Great Lakes Paleo-Indian. Eastern States Archaeological Federation, Proceedings of the Annual Meetings, 1978/1979, p. 16. Supplement to Archaeology of Eastern North America 8.

Roosa, W.B., and D.B. Deller
1982 The Parkhill complex and eastern Great Lakes Paleo-Indian. Ontario Archaeology 37:3–15.

Sanders, T.N.
1983 The Manufacturing of Chipped Stone Tools at a Paleo-Indian Site in Kentucky. Unpublished MA thesis, Dept. of Anthropology, University of Kentucky, Lexington.
1989 The Adams site: a Paleoindian manufacturing and habitation site in Christian county, Kentucky. *In* Paleoindian and Archaic Research in Kentucky, C.D. Hockensmith, D. Pollack, and T.N. Sanders, eds., pp. 1–24. Kentucky Heritage Council, Lexington.

Shafer, H.
1970 Notes on Uniface Retouch Technology. American Antiquity 35(4):480–87.

Sheppard, P.J.
1977 An Archaeological Survey of the Niagara Escarpment Southwest of Collingwood, Ontario (1977). Report on File, Ministry of Citizenship and Culture, Toronto, Ontario.

Shott, M.J.
1986 Settlement Mobility and Technological Organization Among Great Lakes Paleo-Indian Foragers. PhD dissertation, University of Michigan. University Microfilms, Ann Arbor.
1989 Technological organization in Great Lakes Paleoindian assemblages. *In* Eastern Paleoindian Lithic Resource Use, C.J. Ellis and J.C. Lothrop, eds., pp. 221–37. Boulder: Westview Press.

Simons, D.L., M. Shott and H.T. Wright
1984 The Gainey site: variability in a Great Lakes Paleo-Indian assemblage. Archaeology of Eastern North America 12:266–79.
1987 Paleoindian research in Michigan: current status of the Gainey and Leavitt projects. Current Research in the Pleistocene 4:27–30.

de Sonneville-Bordes, D., and J. Perrot
1954 Lexique typologique du Paleolithique Superier; outillage lithique: I grattoirs-II outlils solutreens. Bulletin de la Societe Prehistorique Francaise 7:327–34.

Spence, M., and W. Fox
1986 The Early Woodland occupations of Southern Ontario. *In* Early Woodland Archaeology, K. Farnsworth and T. Emerson, eds., pp. 4–46. Center for American Archaeology, Kampsville Seminars in Archaeology, No. 2.

Spiess, A., and D. Wilson
1987 Michaud: A Paleo-Indian Site in the New England-Maritimes Region. Occasional Publications in Maine Archaeology, No. 6.

Spiess, A., M. Curran and J. Grimes
1985 Caribou (*Rangifer tarandus L.*) bones from New England Paleo Indian sites. North American Archaeologist 6(2):145–59.

Stewart, A.
1984 The Zander site: Paleo-Indian occupation of the southern Holland Marsh region of Ontario. Ontario Archaeology 42:45–79.

Storck, P.L.
1978 Some recent developments in the search for early man in Ontario. Ontario Archaeology 29:3–16.
1979 A Report on the Banting and Hussey Sites: Two Paleo-Indian Campsites in Simcoe County, Southern Ontario. National Museum of Canada, Archaeological Survey of Canada, Mercury Series, Paper No. 93.
1981 A "Behind the scenes" view of fieldwork or The coming of age of a research programme. Royal Ontario Museum, Archaeological Newsletter, New Series No. 188.
1982 Paleo-Indian settlement patterns associated with the strandline of Glacial Lake Algonquin in southcentral Ontario. Canadian Journal of Archaeology 6:1–31.
1983 The Fisher site, fluting techniques, and early Paleo-Indian cultural relationships. Archaeology of Eastern North America 11:80–97.
1984a Research into the Paleo-Indian occupations of Ontario: a review. Ontario Archaeology 41:3–28.
1984b Glacial Lake Algonquin and Early Paleo-Indian settlement in southcentral Ontario. Archaeology of Eastern North America 12:286–98.
1988 Recent excavations at the Udora site: A Gainey/Clovis occupation site in southern Ontario. Current Research in the Pleistocene 5:23–24.

Storck, P.L., and P. Von Bitter
1981 The search changes direction: Ice Age man in Ontario. Rotunda 14(3):28–36.
1989 The geological age and occurrence of Fossil Hill formation chert: implications for Early Paleoindian settlement patterns. *In* Eastern Paleoindian Lithic Resource Use, C.J. Ellis and J. Lothrop, eds., pp. 165–89. Boulder: Westview Press.

Tixier, J.
1974 Glossary for the description of stone tools with special reference to the Epipaleolithic of the Maghreb. Newsletter of Lithic Technology, Special Publication No. 1.

Tunnell, C.
1975 Fluted Projectile Point Production as Revealed by Lithic Specimens from the Adair-Steadman Site in Northwest Texas. Texas Historical Commission, Office of the State Archaeologist, Special Report No. 18.

Turnbull, C.
1974 The second fluted point from New Brunswick. Man in the Northeast 7:109–10.

Voss, J.
1977 The Barnes site: functional and stylistic variability in a small Paleo-Indian assemblage. Midcontinental Journal of Archaeology 2(2):283–305.

Walker, P.L.
1978 Butchering and stone tool function. American Antiquity 43(4):710–15.

Webb, W.S.
1950 The Carlson Annis Mound, Site 5, Butler County, Kentucky. University of Kentucky, Reports in Anthropology, 7(4).
1951 The Parrish Village Site, Site 45, Hopkins County, Kentucky. University of Kentucky, Reports in Anthropology 7(6).

White, A.
1963 Analytic description of the chipped stone industry from the Snyders site, Calhoun county, Illinois. Anthropological Papers, Museum of Anthropology, University of Michigan No. 19:1–70.

Wilmsen, E.
1970 Lithic Analysis and Cultural Inference: A Paleo-Indian Case. Anthropological Papers, University of Arizona No. 16.

Wilmsen, E., and F.H.H. Roberts Jr.
1978 Lindenmeier, 1934–1974: Concluding Report on Investigations. Smithsonian Contributions to Anthropology No. 24.

Winn, C.F.
1977 Vegetational History and Geochronology of Several Sites in South Southwestern Ontario with Discussion on Mastodon Extinction in Southern Ontario. Unpublished MSc thesis, Department of Geological Sciences, Brock University, St. Catharines, Ontario.

Witthoft, J.
1952 A Paleo-Indian site in eastern Pennsylvania: an early hunting culture. Proceedings of the American Philosophical Society 96(4):464–95.

Wright, H.T.
1981a Technology and Economy: Recent Research on Early Hunters of the Central Great Lakes. Paper presented at the 46th Annual Meeting of the Society for American Archaeology, San Diego, California.
1981b A fluted point from the Muskegon Valley. Michigan Archaeologist 27(3–4):87–91.

Wright, H.T., and W.B. Roosa
1966 The Barnes site: a fluted point assemblage from the Great Lakes Region. American Antiquity 31(6):850–60.

Appendix A

Attributes and Variables

In this appendix, the attributes and variables recorded and their method of measurement are summarized. The terminology used to describe a blank or unifacial tool is given on Figure A-1.

GENERAL VARIABLES

1) **Weight**. This was taken to the nearest .01 g using an electronic scale.
2) **Maximum length**. This and other continuous size variables were taken to the nearest .1 mm. Length is the maximum dimension of the item parallel to the axis of percussion (White 1963:12).
3) **Maximum width**. This is the maximum dimension of the flake along a line at right-angles to the axis of percussion.
4) **Maximum thickness**. This is the distance between the dorsal and surfaces beyond the bulbar area (Wilmsen 1970:19) unless the bulb is flat or absent. In that case, the dimension is simply the maximum thickness beyond the platform.

PROXIMAL END

1) **Platform length**. If present, the length of the striking platform between the lateral edges was recorded (Wilmsen 1970:15).
2) **Platform width**. This refers to the maximum width of the platform between the dorsal and ventral surfaces.
3) **Platform lip**. Presence or absence of platform lipping was noted.
4) **Platform angle**. Using a contact goniometer, this angle was measured to the nearest 5° increment. It is the angle between the major platform surface and the main dorsal flake surface in longitudinal section at the proximal end.
5) **Platform surface or preparation**
Several discrete forms of platform surface or preparation were noted, derived largely from Frison and Bradley (1980:27–30). Some of these can co-occur on the same tool.
    a) *Cortical*
    b) *Faceted*. The platform surface exhibits several small flake scars from purposeful platform preparation.
    c) *Ground*. The juncture of the dorsal flake surface and the platform has been dulled to thicken and thus, strengthen, the core edge for the detaching blow.
    d) *Isolated*. The platform outline when viewed from the dorsal surface in plan, has been "pointed" or made to project from the core edge in order to focus the force of the detaching blow at a specific point
    e) *Reduced*. The dorsal surface of the flake at the platform end has been trimmed to remove platform overhang on the core prior to the flake's removal.
    f) *Flat*. The surface of the platform lacks cortex and flake scars.

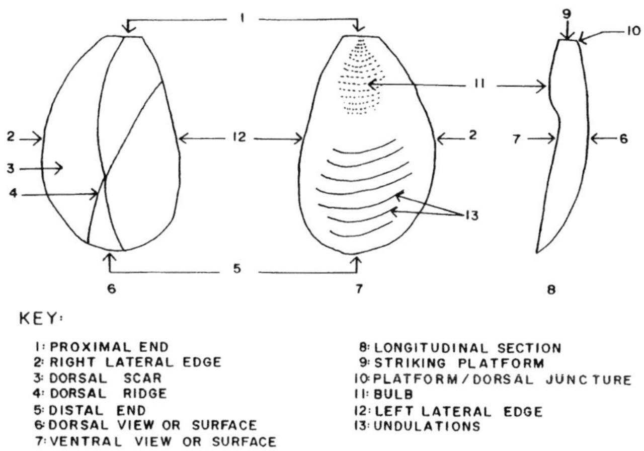

Figure A-1. Flake landmarks.

DORSAL SURFACE

1) **Dorsal scars**. The number of dorsal scars indicating previous removals from the core was recorded. Platform trimming ("reduced") scars are excluded from the counts.
2) **Dorsal scar orientation**
The orientation of the dorsal scars versus the orientation of the flake under examination was recorded.
    a) *Parallel-unidirectional*. Previous removals originated at the proximal or platform end of the flake being examined and paralleled its long axis.
    b) *Parallel-bidirectional*. Previous flake detachments originated at both ends of the flake and paralleled its long

axis. Essentially, two opposing platforms were used to derive flakes from the core.

c) *Convergent-unidirectional*. The flake scars originated at the same platform end but converge towards the axis of percussion of the flake.

d) *Convergent-bidirectional*. The scars originated at both ends and converged towards the axis of percussion.

e) *Transverse*. Some or all of the scars indicate previous removals were at right-angles to the long axis of the flake under examination.

f) *Flat*. The dorsal surface is partially or wholly unflaked.

VENTRAL SURFACE

1) **Bulb type**
   a) *Flat or diffuse*. The bulb is poorly defined and does not project from the ventral flake surface.
   b) *Moderate*. The bulb is well-defined.
   c) *Pronounced*. The bulb is well-defined and projects markedly from the ventral surface.
2) **Other**
   a) *Undulations* (Cotterell and Kamminga 1979:108).
   b) *Smooth*. Undulations are lacking.

TRANSVERSE SECTION

1) **Core facet angle**. This was recorded to the nearest 5° increment using a contact goniometer on all flakes with triangular, offset triangular or wedge-shaped transverse sections. It is the angle between the two major dorsal facets (see Judge 1973:146–148).

LONGITUDINAL VIEW

1) **Curvature**. The curvature of a flake was recorded. It was measured by matching the tool to a series of concentric circles drawn at 1 cm intervals from 1 to 15 cm diameters. Those items which fit a circle greater than about 13 cm are essentially flat while those under ca. 9 mm have pronounced curvature.
2) **Curvature placement**
   a) *Distal*. The curvature is greatest near the distal flake end.
   b) *Symmetrical*. The curvature is well-centered and gradual throughout the flake's length.
   c) *Proximal*. The curvature is pronounced towards the proximal flake end.

OTHER CHARACTERISTICS

1) **Distal termination** (after Cotterell and Kamminga 1979: Fig. 105).
   a) *Feathered*
   b) *Hinge or step*
2) **Lateral edge orientation** (after White 1963:9).
   a) *Parallel*
   b) *Contracting*
   c) *Expanding*. The lateral edges expand from the platform in plan view. The degree of divergence was measured to the nearest 5° increment (Fig. A-2B).
3) **Banding orientation**. If on banded Collingwood chert, the orientation of the bands versus the flake's axis of percussion was recorded to the nearest 5° increment on a 0–90° scale (Fig. A-2A).

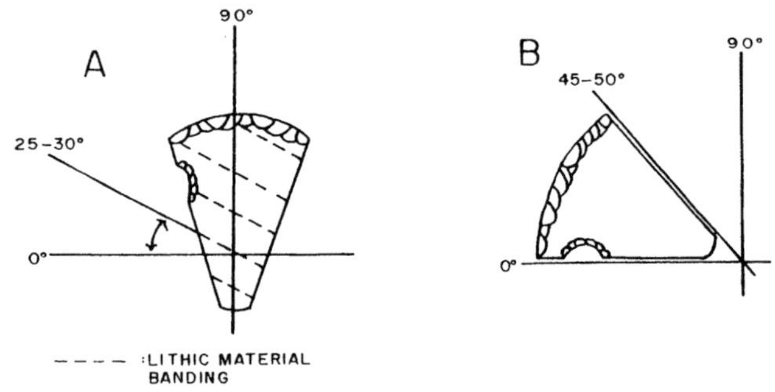

Figure A-2. Measurement of banding orientation (A) and lateral edge divergence (B) on unifacial tool blanks.

RETOUCH CHARACTERISTICS

A number of characteristics were recorded for the retouched edges on unifacial tools.

1) **Retouched surface**
   a) *Normal or dorsal*. The retouch scars occur on the dorsal surface of the flake blank.
   b) *Inverse or ventral*. The retouch scars appear on the ventral surface of the flake blank.
   c) *Bifacial*. The retouch scars appear on both surfaces along the same segment of an edge.
2) **Retouch character**
   a) *Continuous*. The retouch scars occur continuously and overlapping along a tool edge.
   b) *Discontinuous*
3) **Retouch length**
   a) *Fine*. The retouch scar's surfaces do not extend more than 2.5 mm from the edge.
   b) *Extensive*. The retouch scar length extends beyond 2.5 mm from the edge. The maximum length of such retouch was recorded.
4) **Retouched edge length**. This is simply the length in a straight line or axis of the retouched edge. On end scrapers it is referred to as bit width.
5) **Retouched edge depth**. This is the depth of convex working edges from the apex of the retouched edge to the retouched edge length axis. On end scraper bits it is referred to as bit depth.
6) **Retouched edge thickness**. This is the maximum thickness of the retouched edge from the distal end of flake scars to the "underside" of the tool. It is measured as if the edge angle was 90° (i.e., at right angles to the tool body in profile view) and is referred to as bit thickness on end scrapers.
7) **Retouched edge angle**. An attempt was made, following the method of Odell (1979:340) to measure the angle between the retouched surface and the tool's "underside" in profile at a consistent point (2.5 mm) from the edge/underside juncture. A contact goniometer was employed to take this measurement. Since the angle can vary along an edge, the range rather than exact measurements along an edge was recorded. Moreover, the smallest increment to which the edge angle was recorded was 5°.

Appendix B

Data on Fluted Points and Fluted Preforms, Thedford II Site

FC#	Type	Weight (gms)	Length	Width	Thickness	Basal Concavity Depth	Basal Width	Material Type
157+	Pt	—	—	—	—	—	—	B
358+	Pt	—	—	27.2	5.4	—	—	B
574+	Pt	13.8	88.0	26.5	5.9	—	—	B
499	Pt	—	—	—	—	3.4	19.2	C
535	Pt	—	—	—	—	3.0	—	C
569	Pt	21.7	105.0	27.0	7.5	5.0	19.5	C
580	Pt	—	—	—	5.2	3.3	18.3	B
536	Pt	13.0	90.5	25.0	5.5	—	—	B
369	Pt	—	—	22.0	7.0	3.1	16.5	C
80+	Pr[a]	2.5	32.1	16.7	3.6	1.85	16.1	C
94+	Pr[b]	4.6	40.1	22.1	5.1	2.5	18.4	C
502+	Pr	—	65.1	24.8	6.2	6.1	—	C
543+	Pr	—	—	28.6	6.0	4.7	—	B
547	Pr	—	—	—	5.1	—	—	C
563	Pr	—	—	—	—	4.5	—	C
537+	Pr	—	—	25.2	8.3	4.2	19.4	C
480+	Pr	13.2	92.0	26.0	6.2	—	—	B
501	Pr	—	63.7	25.0	6.5	3.7	18.5	C
191	Pr	—	—	—	—	—	—	C

FC# = field catalog number; Type: Pt = point; Pr = preform; Material Type: C = Collingwood; B = Bayport
[a]Small atypical preform.
[b]Preform with refitted outre passe tip—measurements are given for small pointed basal section alone.

FC#	# of Flutes per Face	Flute Width			Flute Length			Face Angle	Banding Orientation
		Obverse	Reverse		Obverse	Reverse			
		1	1	2	1	1	2		
157+	1-1	14.2	—	—	—	—	—	96.5°	N/A
358+	1-1	16.1	16.4	—	—	—	—	—	N/A
574+	1-1	11.0	17.1	—	63.5	72.7	—	96.0	N/A
499	1-2	14.6	—	—	—	—	—	98.0	0-5°
535	1-1	11.0	10.2	—	—	—	—	98.0	1-5°
569	1-1	13.0	14.5	—	82.5	53.0	—	97.25	mottled
580	1-2	—	—	—	—	—	—	96.0	N/A
536	1-2	12.5	12.5	5.5	70.0	62.5	37.1	96.0	N/A
369	1-1	8.0	6.0	—	27.0	19.0	—	98.0	0-5
80+	1-1	5.6	8.0	—	22.1	—	—	N/A	0-5
94+	1-1	13.3	11.0	—	34.9	23.6	—	N/A	0-5
502+	1-1	11.4	11.2	—	43.3	45.3	—	N/A	0-5
543+	1-1	15.1	14.3	—	33.7	—	—	N/A	N/A
547	1-1	17.0	11.5	—	—	—	—	N/A	30-35
563	1-1	14.6	10.3	—	—	—	—	N/A	0-5
537+	1-2	12.6	10.4	6.8	—	—	—	N/A	mottled
480+	1-1	16.4	14.2	—	—	—	—	N/A	N/A
501+	1-1	12.5	7.0	—	42.7	33.0	—	N/A	0-5
191	1-?	14.0	—	—	—	—	—	N/A	10-15

N/A = not applicable
Note: FC#'s 358+, 574+, 499, 569, 502+, 543+, 563, 537+, and 501 are probably part of the cache in Area A-northeast.

Appendix C

Other Fluted Points Examined

#	Type	Reference #	Collection	Figure and #	Material	Length	Width	Thickness	Basal Width	Basal Concavity
1	G	C#979-9-734-783	MIA	32-8	C	39.7	23.2	7.0	23.2	7.4
2	G	Uniondale site	Private	32-12	C	51.0	26.0	8.0	26.0	8.0
3	G	None	Private	32-2	0	-	27.4	6.0	24.5	5.3
4	G	C#4304	MIA	—	0	33.6	24.1	6.1	24.1	6.6
5	G	C#949-215-196	ROM	—	C	63.1	24.4	6.6	22.0	4.2
6	G	None	Private	32-11	B	—	30.0	6.0	28.0	5.0
7	G	Frank Wood Collection, C#9347	McM	32-18	0	63.1	28.1	7.6	27.1	5.0
8	G	Garrad 1971:#13	MIA	32-9	UM	—	25.2	7.2	25.2	5.2
9	G	Garrad 1971:#15; C#6	MIA	32-6	O	—	26.3	7.2	26.3	5.1
10	G	Garrad 1971:#20; C#708	Private	32-5	UM	94.5	27.5	8.5	27.5	3.5
11	G	Garrad 1971:#27; C#7	MIA	32-10	C	64.7	25.6	6.3	22.7	4.4
12	G	Garrad 1971:#33; C#903-9-1; #25810	ROM	32-15	U	56.5	24.7	6.6	24.9	5.2
13	G	Garrad 1971:#35; C#1834	MCC	32-17	O	68.7	25.5	8.0	25.5	6.3
14	B	Garrad 1971:#14; C#3466	MIA	33-4	SC	75.5	25.6	7.3	—	2.0
15	B	Garrad 1971:#34; C#GRDSAS, #1064	MCC	33-8	O	53.2	21.3	4.8	—	5.2
16	B	Schoefield site	Private	2-6	O	51.5	22.5	7.0	21.0	3.0
17	B	None	Private	Near Ann Arbor, Michigan	U	49.5	21.0	5.0	16.0	2.5
18	B	Mullin site	Private	33-5	B	—	—	4.7	19.1	3.5
19	B	McLeod site	UOW	2-3	C	—	19.0	7.0	—	—
20	B	Garrad 1971:#28; C#410	MIA	—	O	—	19.5	5.0	19.5	3.6
21	B	None	Private	—	SC	—	26.5	6.3	—	—
22	B	Babula site	MCC	33-3	UM(?)	—	—	6.8	—	4.8
23	B	Wight site	Private	2-2	SC	—	23.2	6.0	—	—
24	C	Garrad 1971:#46; C#921-10-24-39702	ROM	34-12	C	44.0	25.7	4.1	17.2	1.4
25	C	Tee Creek site	Private	34-5	U	—	—	4.0	17.5	1.5
26	U(Cu?)	Garrad 1971:#8; C#3446	MIA	—	O	—	22.5	8.3	—	4.0
27	U(B?)	Kovacs site	Private	—	O	40.6	20.2	3.5	19.2	5.0
28	U	Weed site	Private	—	O	45.5	21.5	6.0	21.5	3.5
29	U(E?)	Deller 1979:#41	Private	—	O	—	28.5	6.0	28.5	3.0
30	U(C?)	C#J8445	MIA	—	O	—	27.0	5.6	—	—
31	U	C#J4111	MIA	—	O	42.3	18.6	5.4	18.6	4.3
32	U(G?)	Garrad 1971:#47; C#891-6-2	ROM	—	O	—	30.0	7.7	—	—
33	U(G?)	Weed site	Private	—	UM	—	—	—	26.5	2.5
34	U	C#27059	ROM	—	O	50.1	24.5	4.6	24.5	3.6
35	U(G?)	Garrad 1971:#26; C#3445	MIA	—	C	—	25.5	8.2	—	—

#	Face-Angle	Banding Orientation	Flutes per Face	Barnes Finishing	Grinding Left	Length Right	Fish-tail	Flute Length and Width Obverse (1) L	W	(2) L	W	Reverse (1) L	W	(2) L	W	(3) L	W
1	88.5;91.0	85-90°	1-0	Y	20.0	17.8	N	18.6*	14.4	—	—	—	—	—	—	—	—
2	89.0;87.0	75-80	1-1	N	20.5	17.5	N	19.5	15.5	—	—	17.5*	14.0	—	—	—	—
3	90.0;90.0	N/A	1-1	Y	21.0	21.9	N	—	11.2	—	—	—	14.4	—	—	—	—
4	89.0;88.0	N/A	1-1	Y	20.0	19.5	N	19.5	15.0	—	—	18.6	11.5	—	—	—	—
5	92.0;93.0	5-10	1-1	Y	24.4	22.2	N	37.7	13.0	—	—	23.6*	12.5	—	—	—	—
6	91.0;92.0	N/A	1-2	Y	N/R	N/R	N	—	18.0	—	—	—	15.0	—	—	—	—
7	93.0;92.0	N/A	1-3	Y	23.4	23.0	S	29.4*	11.3	—	—	34.2*	—	23.6*	—	25.9*	8.4
8	91.0;89.0	N/A	1-1	Y	19.4	25.2	S	34.1	11.5	—	—	24.2*	12.6	—	—	—	—
9	86.5;96.0	N/A	1-1	Y	23.1	21.2	N	25.8*	12.4	—	—	24.6*	12.6	—	—	—	—
10	90.0;91.0	N/A	1-2	?	22.0	26.5	N	20.0*	13.5	—	—	30.5*	12.5	30.5*	—	—	—
11	91.5;94.0	5-10	1-1	Y	22.0	22.0	N	46.4	10.8	—	—	47.5	12.7	—	—	—	—
12	92.5;93.0	N/A	2-0	Y	21.5	19.1	S	27.0*	11.0	18.3*	16.6	—	—	—	—	—	—
13	88.5;87.5	N/A	2-0	Y	23.5	22.2	N	24.0*	—	21.1*	7.2	—	—	—	—	—	—
14	95.5;97.5	N/A	1-2	Y	19.0	—	S	61.3*	13.5	—	—	66.3*	14.6	—	—	—	—
15	96.5;—	N/A	1-1	Y	17.0	14.5	Y	25.7*	11.5	—	—	29.7*	7.8	—	—	—	—
16	95.0;94.5	N/A	1-1	Y	17.0	14.5	Y	20.0*	8.0	—	—	15.5*	7.5	—	—	—	—
17	99.0;99.0	N/A	1-1	Y	18.5	19.5	S	30.0	12.5	—	—	14.5	12.0	—	—	—	—
18	97.0;98.0	N/A	1-1	N	21.4	23.5	Y	—	11.5	—	9.4	—	—	—	—	—	—
19	—;—	N/A	1-2	?	?	?	?	—	9.6	—	7.2	29.3*	7.6	—	—	—	—
20	92.5;91.0	N/A	1-0	Y	?	?	S	29.4*	9.6	—	—	—	—	—	—	—	—
21	—;—	N/A	1-1	?	?	?	?	—	14.5	—	—	—	12.5	—	—	—	—
22	97.5;—	N/A	1-2	Y	?	17.2	N	—	13.6	—	—	—	8.5	—	—	—	—
23	—;—	N/A	1-1	?	?	?	?	—	12.1	—	—	—	12.0	—	—	—	—
24	102;101.0	45-50	2-0	N	20.5	22.2	N	32.0	12.7	25.0*	—	—	—	—	—	—	—
25	101;101.0	N/A	2-3	N	?	?	N	—	7.8	—	—	20.0	6.5	23.0	11.5	—	—
26	94.0; 96.0	N/A	1-1	Y	31.0	?	Y	28.0*	10.1	—	—	63.3*	13.4	—	—	—	—
27	97.5;96.0	N/A	1-2	Y	12.1	13.1	Y	25.5*	8.0	—	—	19.5*	8.5	16.5*	7.5	—	—
28	91.0;89.0	N/A	1-1	Y	23.5	23.5	Y	30.5*	9.0	—	—	21.5*	9.5	—	—	—	—
29	88.0; 88.0	40-45	1-1	?	?	?	S	18.0*	15.0	—	—	16.5*	16.5	—	—	—	—
30	—;—	N/A	1-?	?	?	?	?	—	14.2	—	—	—	—	—	—	—	—
31	91.0;89.0	N/A	1-1	Y	16.5	16.2	N	29.0	8.4	—	—	21.0	8.7	—	—	—	—
32	—;—	N/A	1-1	?	?	?	?	—	19.6	—	—	—	16.5	—	—	—	—
33	87.0;91.0	N/A	1-0	Y	18.0	?	S	—	—	—	—	—	—	—	—	—	—
34	91.5;92.0	N/A	2-2	N	20.5	18.8	S	41.1	8.1	41.1	—	22.5*	13.0	—	—	—	—
35	—;—	0-5	1-2	?	?	?	?	—	11.1	—	—	—	11.1	—	6.0	—	—

KEY

Type
G= Gainey; B= Barnes; C= Crowfield; U= unknown; Cu= Cumberland

Reference
Includes, where applicable, site name, literature reference, catalog number written on artifact (C#)

Collection
MIA = Museum of Indian Archaeology, London, Ontario
ROM = Royal Ontario Museum
McM = Dept. of Anthropology, McMaster University, Hamilton, Ontario
MCC = Ministry of Citizenship and Culture, London, Ontario
UOW = Dept. of Anthropology, University of Waterloo, Waterloo, Ontario

Figure and #
refers to figure # and location # in text

Material
C = Collingwood; SC = soft Collingwood; O = Onondaga;
B = Bayport; UM = Upper Mercer, Ohio; U = unknown chert

Barnes Finishing
Y = yes; N = no

Fishtail
Y = yes; N = no

Other
N/A = not applicable
N/R = not recorded
Flute Length: asterisked items are original flute lengths unaffected or not reduced in length by resharpening

APPENDIX D

Cross Reference of Field Catalog Numbers and Illustrations

PART I—By FC

F.C. #	Figure #	F.C.#	Figure #	F.C.#	Figure #
19	51c, 55b	53/247	59h	150	60g
20	55e	55	50e, 51b	153/235	59d
21	52a	57/133	61h	155/220	56c
22	55d	64	61d	157/540	20h
23	52f	65	52d	158	52e
24	55f	67	59i	170	56i
25	50f, 51a	68+	56a	175	54e
27/28	62a	74	61c	177	62b
29	23e, 43b	75+	56d	178+	59g
30	43c	76/96	56l, 57c	182	60j
31	61e	77	60f	183+	61b
32	59e	80/189	22i	187/396	61i
33	43d	81	60c	191/616	22h
34	59f	84/113	26d	192	60d
35	26c	90	50d	194	57b, 60h
36	23d, 26b	92/394	58d	196	61e
37	42b	94+	22e, 27,28	204	60e
43	44	99	61d	207/245	52h
44	53e	117	60l	216	60b
45	53h	124+	60o	224	61g
46	57d, 59c	126	55a	232	52g
50+	60k	134/195	54f	236	59a
52/271	55g	149	60a	239	61f

240	53c	368/501	22c, 23b	548	51e, 54c
249	55h	369	20e	549	50g
254	56k	473	43a	554	58c
255	24e	474	56f	555	56e
256	54h	475	55i	563	22f
257	43e	476+	53f	564	42c
261	52b	478+	22d, 23a	569	20a, 21a
271	50i	480/484	22a	570	26e
275	56j	482	24f	580	20f
276	24h	486	53a	592	23c, 26a
279	58e	489+	20c, 21b	594	50b
283	60i	491	58a	596	20b
285	60n	499	20i	603	58f, 61a
286	61c	503	50c	608	54b
306	59b	508	61f	609	20j
313	24c	510	50h	664	54a
320	42a	516	52c	665	50j
322	61g	517	53b	666	58g
324	55c	527	50a	667	54g
326	53d	528	51d, 53g	668	58b
329	56h	535	20g	681	56g
330+	54d	537+	22b	682/684	56b
351+	24a	542/543	22g		
358/590	20d	545	24d		
359	24b	547	22j, 24g		

PART II—By FIGURE

Figure #	F.C.#	Figure #	F.C. #	Figure #	F.C. #
20a	569	23a	478+	42c	564
20b	596	23b	368+	43a	473
20c	489+	23c	592	43b	29
20d	358/590	23d	36	43c	30
20e	369	23e	29	43d	33
20f	580	24a	351+	43e	257
20g	535	24b	359	44	43
20h	157/540	24c	313	50a	527
20i	499	24d	545	50b	594
20j	609	24e	255	50c	503
21a	569	24f	482	50d	90
21b	489+	24g	547	50e	55
22a	480/484	24h	276	50f	25
22b	537+	26a	592	50g	549
22c	368/501	26b	36	50h	510
22d	478+	26c	35	50i	271
22e	94+	26d	84/113	50j	665
22f	563	26e	570	51a	25
22g	542/543	27	94+	51b	55
22h	191/616	28	94+	51c	19
22i	80/189	42a	320	51d	528
22j	547	42b	37	51e	548

Figure #	F.C.#	Figure #	F.C. #	Figure #	F.C. #
52a	21	54g	667	56l	76/96
52b	261	54h	256	57a	150
52c	516	55a	126	57b	194
52d	65	55b	19	57c	76+
52e	158	55c	324	57d	46
52f	23	55d	22	58a	491
52g	232	55e	20	58b	668
52h	207/245	55f	24	58c	554
53a	486	55g	52/271	58d	92/394
53b	517	55h	249	58e	279
53c	240	55i	475	58f	603
53d	326	56a	68+	58g	666
53e	44	56b	682/684	59a	236
53f	476+	56c	155/220	59b	306
53g	528	56d	75+	59c	46
53h	45	56e	555	59d	153/235
54a	664	56f	474	59e	32
54b	608	56g	681	59f	34
54c	548	56h	329	59g	178+
54d	330+	56i	170	59h	53/247
54e	175	56j	275	59i	67
54f	134/195	56k	254	60a	149

Figure #	F.C. #	Figure #	F.C #
60b	216	61c	286
60c	81	61d	99
60d	192	61e	31
60e	204	61f	508
60f	77	61g	224
60g	150		
60h	194		
60i	283		
60j	182		
60k	50+		
60l	117		
60m	McLeod		
60n	285		
60o	124+		
61a	603		
61b	183+		
61c	74		
61d	64		
61e	196		
61f	239		
61g	322		
61h	57/133		
61i	187/396		
62a	27/28		
62b	177		